TOP 100

ADAM SPENCER'S TOP 100

x

First published in Xoum by Brio Books in 2018
This edition published in 2019

Brio Books Pty Ltd
PO Box Q324, QVB Post Office,
NSW 1230, Australia
briobooks.com.au

ISBN 978-1-922267-02-3 (print)

Cover design, text design and typesetting by Xou Creative, xou.com.au

Author photograph by David Stefanoff
Additional cover photography from Adobe Stock

Printed and bound in China through Asia Pacific Offset Limited

To my mum Elizabeth, the closer you get to 100, the more I realise you are #1

To Ellie and Olivia, 13 and 11 and climbing much too fast

To Yana, age is but a number but no one has ever made me feel so young

Welcome to the *Top 100*

This book is a tribute to the amazing hotness of the numbers 100 to 1.

As we count down these incredible integers, we'll also stop along the way to salute some of their less acknowledged buddies — the various decimals, fractions and the like which all come together to make the splendour of this top 100 countdown.

I'm often asked in radio or TV interviews (or just randomly on the street) why I love numbers and mathematics so much. The more I think about it, the more I gravitate to the following answer. I think the essence of being human, that which separates us from trees, mosquitoes, three-toed pygmy sloths and it seems practically every other form of life, is the extraordinary blob of matter in our skulls — our brains. These collections of 100 billion neurons, give or take, which join up to form trillions upon trillions of synapses and are fuelled by a remarkably complex dance of chemicals and electricity, let us speak, dream and most importantly think. And of all the things we think about, one of the most fundamental questions occurs when we look around us and wonder what the heck is going on. We have so many tools at our disposal to analyse our world — to measure it, understand it, explain it to others and hazard a guess as to why, how or what next. But surely our most powerful tool is mathematics. Numbers, logic, the language of maths and their applications have, I humbly submit, contributed more to answering this most deeply human question than any of our myriad other forms of observation and expression.

That's not to say Picasso's *Les Demoiselles d'Avignon* is not an extraordinary work of art that revolutionised the way the human form was depicted. Nor that JK Rowling's books haven't made the lives of millions of children the richer. Nor that I didn't hug a stranger and shed tears of joy when the

Sydney Swans won the 2005 Grand Final. But compared to what numbers can reveal, all of these ways of being come a distant second. (Again, in my humble opinion!)

For those of us lucky enough to 'get it', mathematics is a wonder in and of itself. It's a place where, as we learn, we acquire more and more tools. A place of elegance and beauty. A place populated by seemingly unanswerable questions, which then succumb to examination, only to throw up even more questions. A place in which the difference between what we expect the world to be and what we measure can be fantastically small.

Take, for example, the electron g-factor. Don't sweat it if you don't know what that is. It's quantum mechanics, after all, and who really understands that? But when it comes to comparing what the maths tells us about this way to measure an electron on an incredibly small scale and what we can actually observe, the degree of the match is astonishing. We know that the electron g-factor is equal to –2.00231930436182 give or take 0.00000000000052. Read that again; give or take 0.00000000000052. That's the equivalent of me asking you to guess how far it is around the Earth and your guess being accurate to within 0.01 of a millimetre.

That is truly beautiful.

We live an in age screaming at us that we need more young people turned on to this beauty. We're constantly told that the next generation needs to be equipped with creative, disciplined and inquiring minds to carry us through the decades of disruption ahead. If this book can do anything to inspire in the young (and not so young) my passion for the beauty of mathematics, I'll be *ridic* happy.

Please enjoy my *Top 100*.

A.S.

With thanks

I cannot take all the credit for this *Top 100*, but I can certainly shoulder all the responsibility for its content. If any of it doesn't make sense or you have any issue with what lies within, I'm only an email away. Reach me at book@adamspencer.com.au.

This year the puzzle-meister Sean Gardiner rose to new heights. Sean, mathematics is the marriage of intellectual creativity and ruthless rigour. You share both in a heady mix which is found in very few. This book, more than any previously, simply could not have happened without you.

To Ruben Meerman, who offered wise counsel when attempting to distil the complex (while not losing its essence), again thanks.

To the army of online bloggers, Tweeters, authors and the general geek-tariat who have inspired so much of the content here, I hope I have attributed fairly when quoting the work of others and offer a genuine apology in the case of any omissions.

To the staff at the New York Museum of Mathematics who don't even realise that on the day I visited you helped inspire me to keep spreading the word ...

Team Brio, it's been another wild ride. Complete with crushing deadlines, an author who didn't keep track of said crushing deadlines, a quick two-week overseas holiday at the most disastrous time imaginable, and a lovely chat with the good people at ABC corporate comms three days out! To Ro-mo for your watchful oversight, albeit from a distance, and to Mac Jonny for your Herculean efforts and skill – sincere thanks. Roy Chen, you surely are the Kendrick Lamar of The Adobe Creative Suite.

And Hardie Grant, great to have you on board – or perhaps to be on board? Regardless of where the board lies ... let's rock!

And let's not forget the shout outs to the awesome guinea pigs who were subjected to my early drafts for feedback: Jarrod Spiga, David Henley, Richard Henley, and Angus MacDonald!

Hit

Let's take it
from the top ...

Big, fat 100!

The website mytechinterviews.com recommends this question for sorting out the nicely nerdy from the genuinely geeky when it comes to job interviews in the tech sector:

'How many zeroes are there in the tail of the number 100 factorial?'

Now, it's early on in the *Top 100*, so perhaps the first thing you're asking is 'what on Earth is a 100 factorial?'. Well, it's a concept you'll see a bit of in this book, so let's walk through it. A factorial of a whole number is the result obtained by multiplying the original number by all the counting numbers less than it. So '6 factorial' is equal to $6 \times 5 \times 4 \times 3 \times 2 \times 1$.

As you will see throughout this countdown, factorials arise all the time in maths problems, especially when we are counting up the different ways we may do something – like draw 3 cards from a deck of 52, or arrange 6 kids in a line.

So 100 factorial, which we write as 100!, is equal to:

$$100 \times 99 \times 98 \times 97 \times 96 \times 95 \times \ldots \times 5 \times 4 \times 3 \times 2 \times 1.$$

I'll tell you now that 100! is a massive number. It's 158 digits long. To put this in perspective, if every atom in our observable universe was itself a copy of our observable universe, 100! would be a pretty good guess for the number of atoms in this 'super universe'.

I'll also tell you that the last several digits of 100! are zeroes. This is what I mean by the 'tail' of the number.

Now, can you tell me how many zeroes there are in the tail of 100 factorial?

The answer's at the back of the book.

100(0000000
0000000000
0000000000
0000000000
0000000000
0000000000
0000000000
0000000000
0000000000
00000000000)

If you've read that bestselling nail-biter of a read *Adam Spencer's World of Numbers,* you won't need Google to tell you that this googly of a number above is none other than a googol, AKA a one followed by 100 zeroes. It's a decently-sized number, I'm sure you'll agree.

But a googolplex is equivalent to 10 raised *to the power of* a googol! That's 1 followed by a googol of zeroes!

Meh, you say. Still not impressed? Well, my extremely hard to impress maths nerd friend, why not try a googolplexian on for size? That's a 1 followed by ... a googolplex of zeroes.

100

Blankety blanks

You like puzzles? Ready to dive on in?

Let's roll!

The rules for these puzzles are pretty simple. Reach the target number by filling in the blanks in each equation.

But hold on. You can't move the operations (+, –, ×, ÷) found between the blanks. Order of operations applies, which means we calculate brackets first, after which × and ÷ trump + and –.

Oh, and a completed equation must incorporate each and every digit from 0 to 9, including the digits in the answer.

Alright, let's kick things off with an example.

$$\bigcirc \times \bigcirc \times \bigcirc - \bigcirc \times \bigcirc - \bigcirc - \bigcirc = 83$$

The solution that uses each of the digits from 1 to 9 is:

$$7 \times 5 \times 4 - 9 \times 6 - 2 - 1 = 83$$

Got it? Then cast your eyes no further than the opposite page. You'll find a few more of these babies throughout the book. Brace yourself — it should come as no surprise that they get harder and harder as we count down to 1.

Blankety blanks (#100)

$$(\bigcirc + 2) \times (\bigcirc + (\bigcirc \times 7 + \bigcirc) \div (\bigcirc + \bigcirc)) = 100$$

$$((\bigcirc \times (\bigcirc - \bigcirc) + \bigcirc) \div 2 + 6) \times (\bigcirc - \bigcirc) = 100$$

$$(\bigcirc + \bigcirc \div (\bigcirc - \bigcirc)) \times (8 - 3) \times (\bigcirc - \bigcirc) = 100$$

$$(\bigcirc \times (\bigcirc - (\bigcirc + \bigcirc) \div \bigcirc) + \bigcirc) \times (3 + 2) = 100$$

$$((\bigcirc - \bigcirc) \times (\bigcirc - \bigcirc) + \bigcirc) \times (3 + 2) + \bigcirc = 100$$

The 0 and 1 are in the answer, so you need to provide the 2, 3, 4, 5, 6, 7, 8 and 9.

I've provided a few hints to ease you in.

Now, all the puzzles are, of course, solveable without hints.

So fear not, my hardcore puzzlers ... you can rest assured that you can still find a unique solution even if you ignore the hints provided!

99.965

I vant a vanta

How dark is dark? How dark can something get? I'm not talking about the sort of jokes that leave you feeling uneasy. I'm talking about the darkest substances on Earth, those which absorb the most light without reflecting anything back.

The good people at Surrey NanoSystems have created a chemical called Vantablack (Vertically Aligned Carbon Nanotube Arrays) which absorbs up to 99.965% of radiation in the visible light spectrum. Ten thousand of these nanotubes of carbon, side to side, would be as wide as a human hair!

That's dark.

But amazingly, it's only a fraction better than the black feathers on a male superb bird of paradise. They absorb 99.95% of light.

Here's how Vantablack actually works.

When a photon hits a Vantablack surface, instead of being reflected, it becomes trapped between the carbon nanotubes. It bounces around in there for a bit before being absorbed and dissipating as heat.

Feral cats have invaded 99.8% of Australia

Check in at number 30 to see what we're doing about this.

99

Debuting at number 99

No biggie, but just quietly, right at the top we're about to debut an entirely new type of puzzle!

Fresh from the mind of maths wiz and good mate Sean Gardiner, allow me to introduce ... the Syndesis.

The fancy name describes the paired connections used in our next puzzle, as well as referencing firing neurons by way of synapsis, and the Greek word for logic. Nifty, eh?

How do they work?

The real beauty is their elegant simplicity.

Enter the numbers 1 to 9 into each 3 × 3 grid so that each circled number is the result of adding, subtracting, multiplying, or dividing the two numbers in the cells it touches.

To ease you into this world-first new puzzle, I've provided part of the answer for you. But don't get too complacent – my generosity will gradually diminish as we count down!

You'll see more of these puzzles throughout the book and – who knows? – maybe elsewhere in the future.

Whatever the case, you saw them here first. A worthy addition to our *Top 100*!

Have fun.

Sean's Syndesis

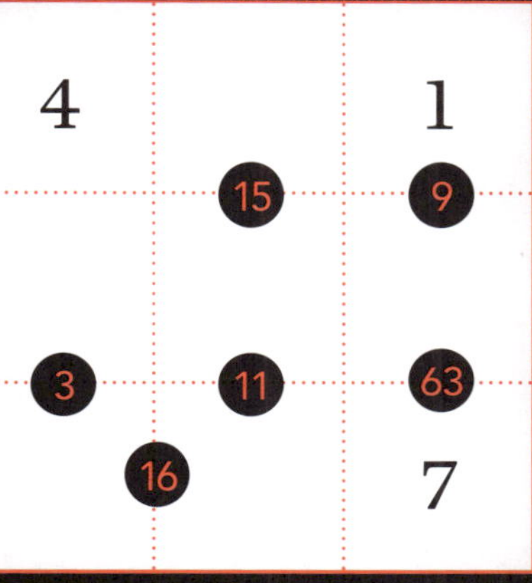

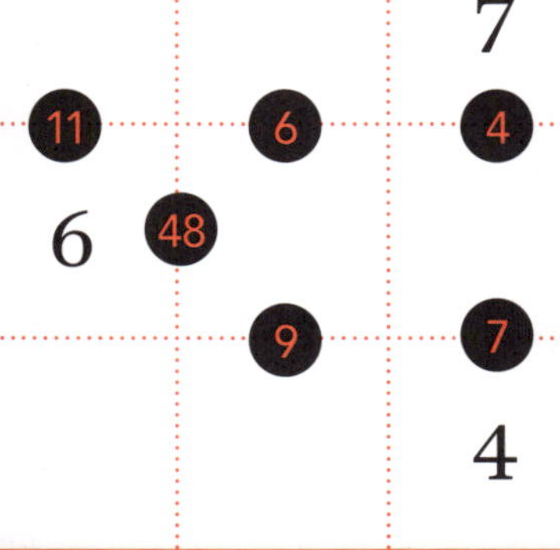

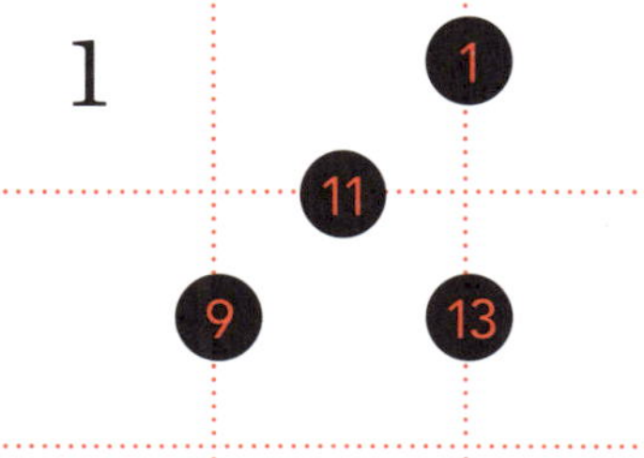

99

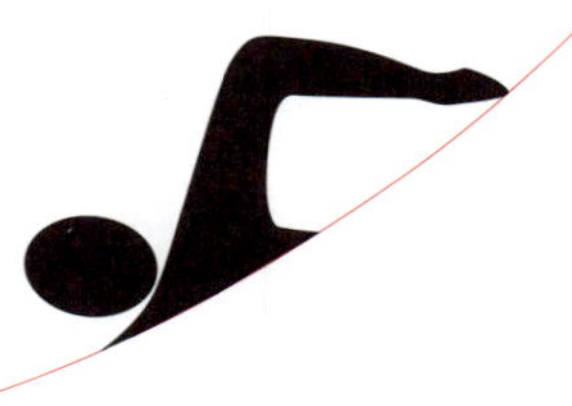

Nonastyle

Setting any world record is impressive, but in February 2018, when 99-year-old George Corones swam 50 metres freestyle in 56.12 seconds, he didn't break the world record for his age group — he demolished it. Being the only person in the race didn't stop the nonagenarian nautilus from shaving 35 seconds off the previous fastest ever effort.

Gude! Oi namo? Ich will klettern du!

According to those very busy elves at @qikipedia, if you were to pick two people from Papua New Guinea at random, there would be a 98.8% chance that they would speak different first languages.

They may speak Tok Pisin, Hiri Motu, Unserdeutsch ... or any one of over 830 languages!

98

Doin' a jig

The number jigsaw, that is!

If you read my last book *The Number Games* (which of *course* you did), you might remember the drill, but just to refresh your memory, here are the rules:

- Arrange each set of 9 tiles into a 3 × 3 square so that adjacent tiles show the same numbers along their touching edges.
- Tiles may need to be rotated, but never reflected.
- Each set of tiles has only one solution, but you might of course end up with a different rotation from me.

So you don't have to cut up your pristine copy of the book (or go crazy copying the pieces yourself), you can find a printout of the pieces in the resources section of my website at adamspencer.com.au/resources.

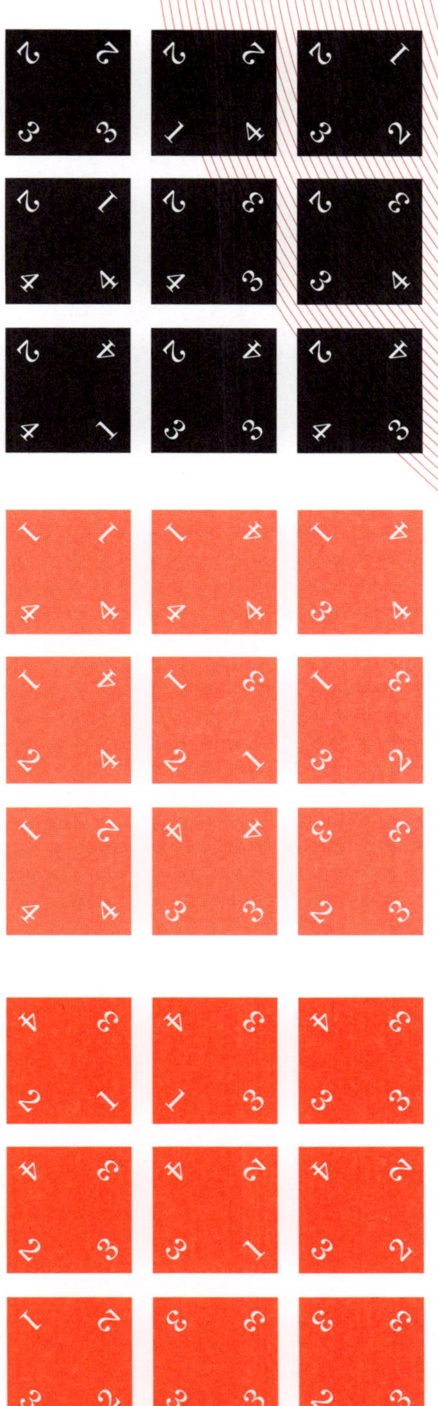

Notice that in each section of 9 tiles, the sum of all the numbers is 98.

Again, thanks to puzzle maestro Sean Gardiner for this cracker.

97

Flight stimulator

Like your flight-related numbers, my geeky friend? Try these Sydney Airport stats on for size:

The T1 International terminal alone handles 14.9 million passengers a year.

The airport also handles 517,000+ tonnes of air freight per year. Approximately 80% of that is carried in the holds of passenger aircraft.

According to an article in *The Sydney Morning Herald*, up to 97% of people who sit the examination for air traffic controllers ... fail.

Now, within the confines of this book I can't expose you to the simulator and make you land a 747 in the fog, but I can replicate the logic puzzles presented by the *SMH* as the first stage of the examination process.

Ready for take-off? Here goes ...

Study the top line of each of the questions on the opposite page. Which of the symbols on the second line – A, B, C, D or E – logically comes next in the pattern?

Answers when we touch down at the end of the book.

Question 1

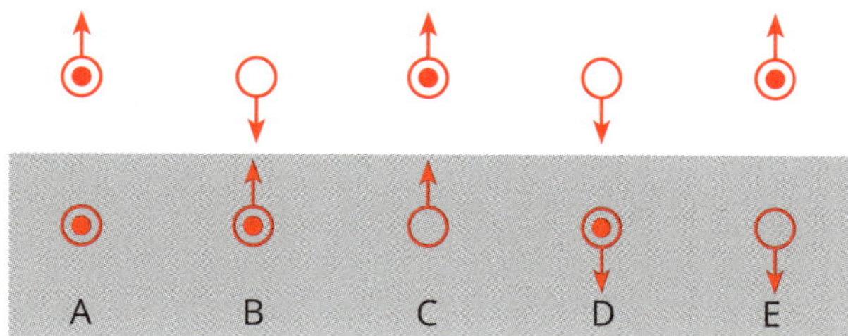

Question 2

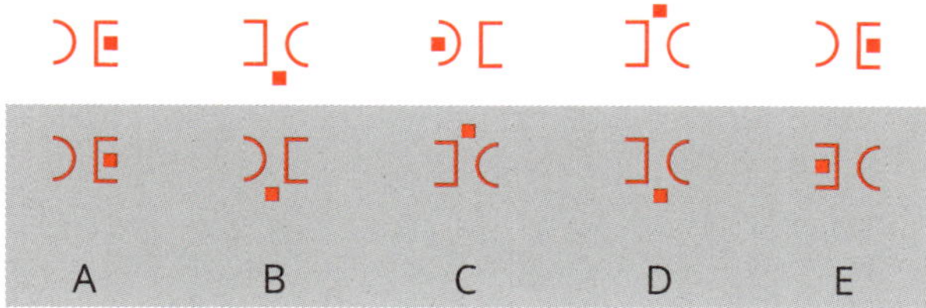

Question 3

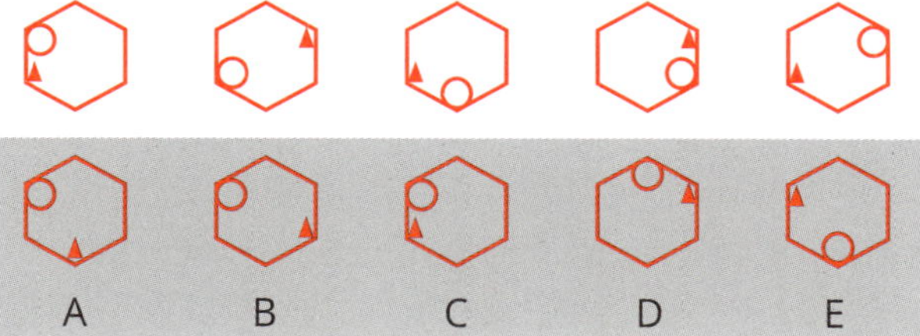

Question 4

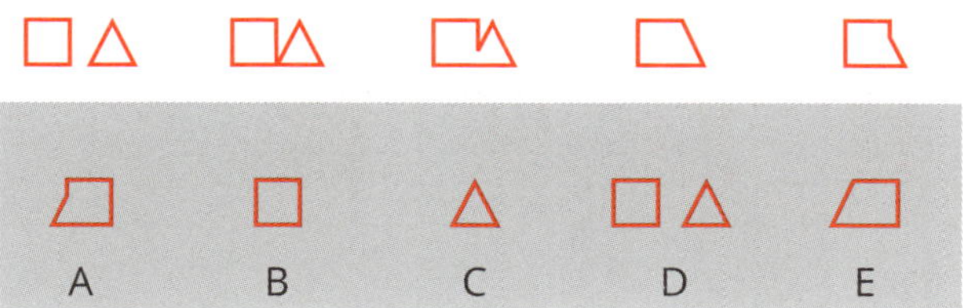

97

New high score!

If you partook in the quest for mathematical survivial which was *The Number Games* (my last book, available where good books are sold, well worth checking out ... sorry) you might recall we met these fun and sometimes fiendish equations.

Well, they're back, baby!

Reach the goal number by creating an equation using the provided numbers and your own choice of operators.

A few house rules for you:

- The numbers should be placed in the white squares.
- Operations (+ – × ÷) are placed in orange squares.
- Order of operations matters, and you can't use brackets!*
- Read the numbers off in order for your final score.
- Your goal is to find the equation that gives the highest score.
- I've given you a few of the operations as a hint to ease you in.

* I'm sure I don't have to remind you that × and ÷ are done before + and – no matter where the symbol occurs.

Let's try an example.

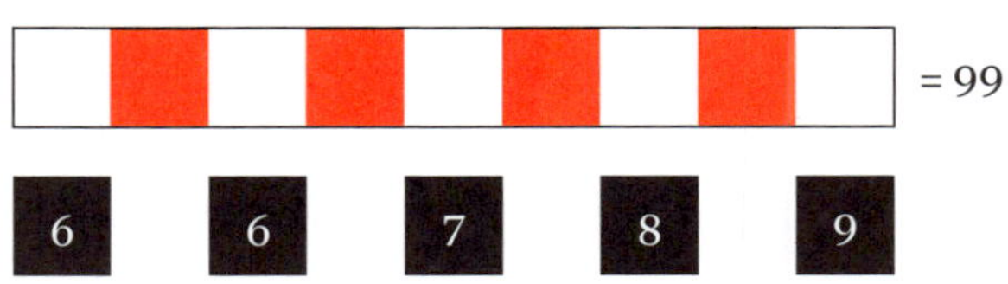

The highest score is 98676 from solution ...

Got it? Have fun!

97

	×		+		×		×		= 97

2 3 5 7 9

	×		×		−		−		= 97

1 3 5 7 7

	×		×		−				= 97

2 6 7 8 9

	×		×		−				= 97

2 4 5 5 6

	×		×		÷				= 97

3 5 6 7 9

96

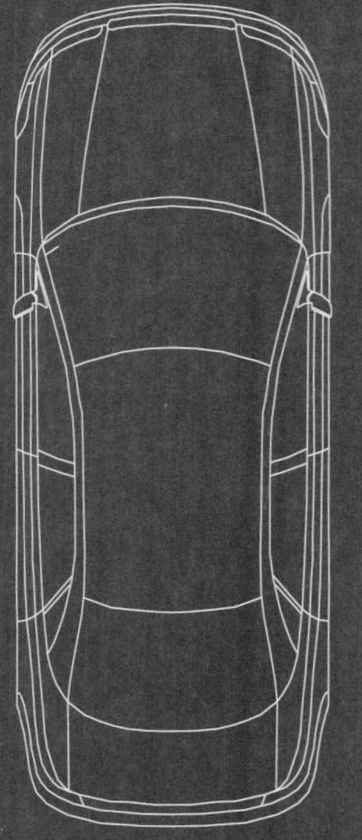

Dude, here's my car!

According to the RAC Foundation, the average car in Britain is parked a whopping 96% of the time. That's made up of 80% being parked at home, 16% being parked somewhere else and just 4% (or less than one hour a day) going around doing car-type stuff.

Pi'd Parseval

On the very first entry in our countdown, I hit you with a piece of mathematical notation that you may never have seen before.

Recall that 6! = 6 × 5 × 4 × 3 × 2 × 1 (pronounced '6 factorial').

Here's another handy example of mathematical notation. This time, a formula that involves the number 96.

$$\sum_{n=0}^{\infty}\left(\frac{1}{(2n+1)^4}\right)=\frac{1}{1^4}+\frac{1}{3^4}+\frac{1}{5^4}+\frac{1}{7^4}+\cdots=\frac{\pi^4}{96}$$

Now, $\pi^4/96$ = roughly 1.014678 (to the first 6 decimal places). How quickly does our series 'converge' to the value $\pi^4/96$?

Let's see how long it takes to match the first 4 decimal places as we terminate the infinite series at $n = 0$, then $n = 1, 2, 3$ and so on.

1.0000, 1.0123, 1.0139, 1.0130, 1.0143, 1.0145, 1.0145 and after 7 terms, that is as far as $1/13^4$ we get the sum to be roughly 1.0146, agreeing with $\pi^4/96$ to the first 4 decimal places.

The perhaps slightly intimidating symbol 'Σ' at the beginning of the equation is the capital letter 'sigma' from the Greek alphabet. It is also used to show that we are taking the sum of a list of terms.

To get the terms of the sum, we calculate $1/(2n+1)^4$ for the values $n = 0$, $n = 1$, $n = 2$, $n = 3$ ... all the way to whatever is on top of the Σ.

In this case, it's infinity. Yep, we're at number 96 and I'm introducing an infinite sum!

96

A hex upon us

Take a look at the hexagonal grids on the opposite page. Your job is to place the 10 given numbers into each one to satisfy all the rules. Which are? Glad you asked!

Each hexagonal cell is touching one or more categories in circles. The number in each cell must belong to all the categories touching it.

Now, fear not – these hexes actually get *easier* as you work your way through the book.

But let's start at the beginning, as we count down to number 1, with this little brain-bender.

Place all the numbers 91 to 100 into each grid.

I've filled out the first one to kick things off, but after that it's just you and your pencil ... and your eraser.

Good luck and have fun!

Oh, and check your answers at the back of the book.

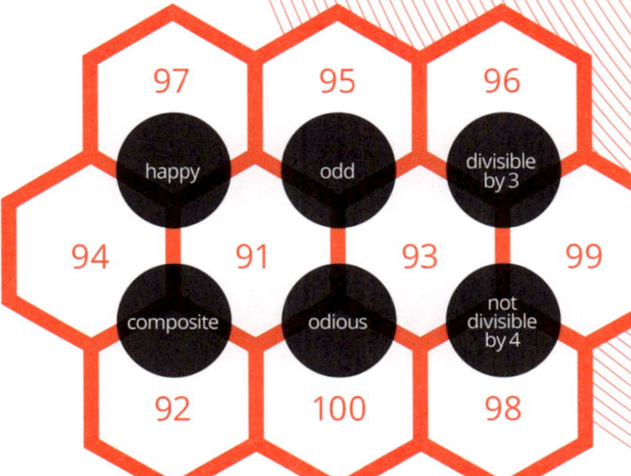

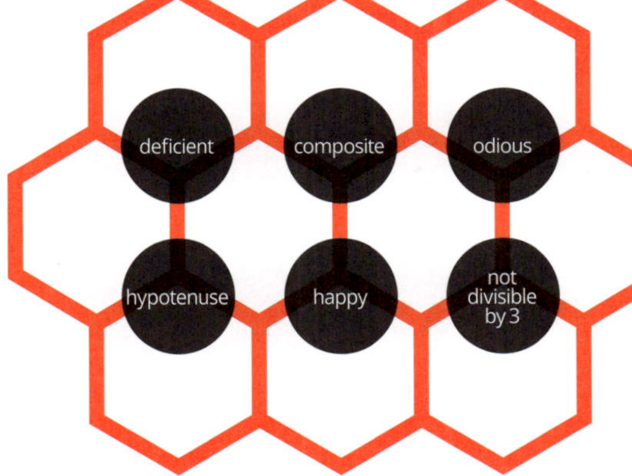

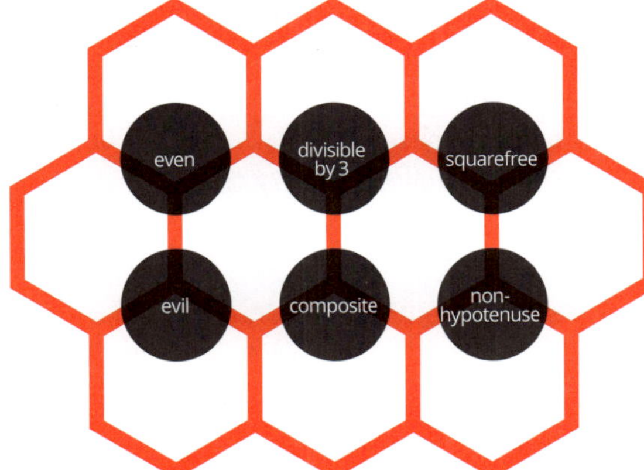

Place all the numbers 91 to 100 into each grid so that each number satisfies the definition in every circle that its grid touches.

So here, 94 is 'happy' and 'composite' while 98 is simply 'not divisible by 4'. But 93 is 'divisible by 3', 'odd', odious' AND 'not divisible by 4'.

Need a hint? If the definitions are a bit tough for you, there is a full list of what these words mean at the end of the book.

Once you've looked up the definitions, if it's still challenging, these hexes are actually easier for smaller numbers — so maybe head to number 6 then work your way back through the hexes (16, 26, 36 and so on).

By the time you get back here, it should all make sense.

95

pq your interest

The number 95 never quite made it as a prime.

It's the first semiprime we encounter travelling down from 100 (and the 34th if we're counting up).

In mathematics, a semiprime (also called biprime or 2-almost prime, or a *pq* number) is a natural number that is the product of two (not necessarily distinct) prime numbers.

For example, 95 is semiprime, and its only factors are 1, 5, 19 and 95.

Why don't you find the others over 90?

Easy peasy.

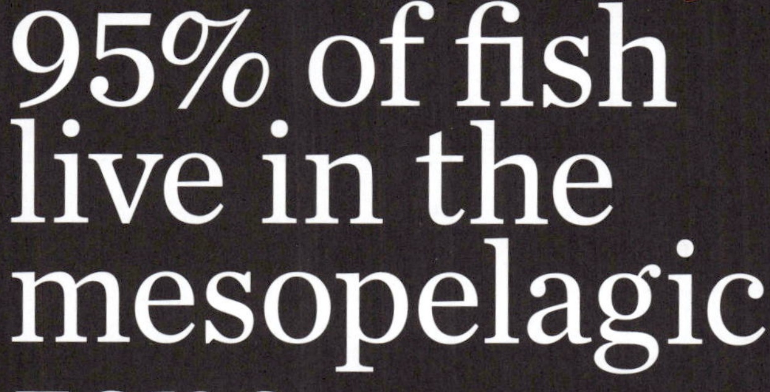

95% of fish live in the mesopelagic zone

That's the layer of the ocean 200–1000 metres underwater.

The temperature ranges across a fish-friendly 4 to 20°C.

95

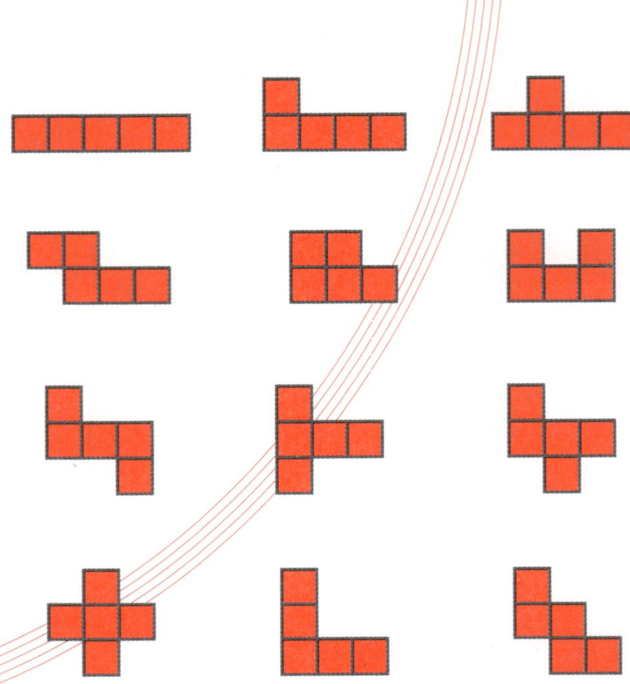

On the tiles

There's more than one way to skin a cat, so they say. Less gruesomely, there's more than one way to tile certain grids with 12 different pentominoes.

In each of these grids, finish tiling it so that it contains one of each pentomino type.

The 12 different pentominoes are shown above.

You may need to rotate and/or even flip some of them to get them to fit.

The total number of hints given per number is equal to the first digit of the chapter number. So there are 9 hints (3 lots of 3 given tiles) for 95, 8 hints for 85 and so forth. Yep, you guessed it: they get harder as we go along.

Obviously I can't give you fewer than 3 hints for the 25, 15, and 5 cases — so just roll with me on that.

Have fun!

Don't forget you can head over to my website (adamspencer.com.au/resources) to download a more cut-uppable copy of this and other puzzles.

94

The Kingdom of Great Britain existed as a country for 94 years

Also known just as 'Great Britain', it was born out of the union of The Kingdom of England and The Kingdom of Scotland on 1 May 1707 and was supplanted on 1 January 1801 by 'The United Kingdom of Great Britain and Ireland'.

94

Gridlock!

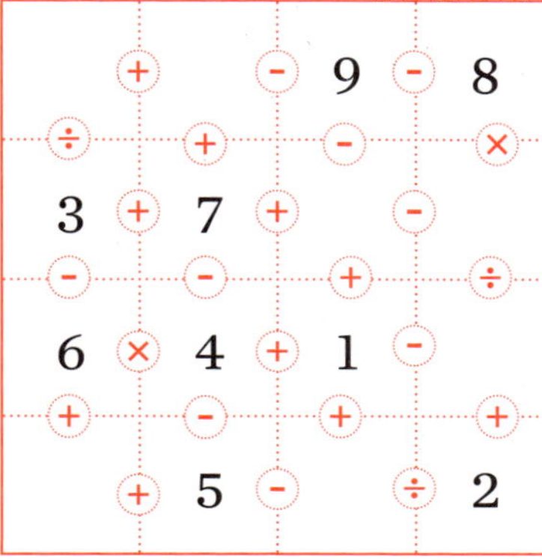

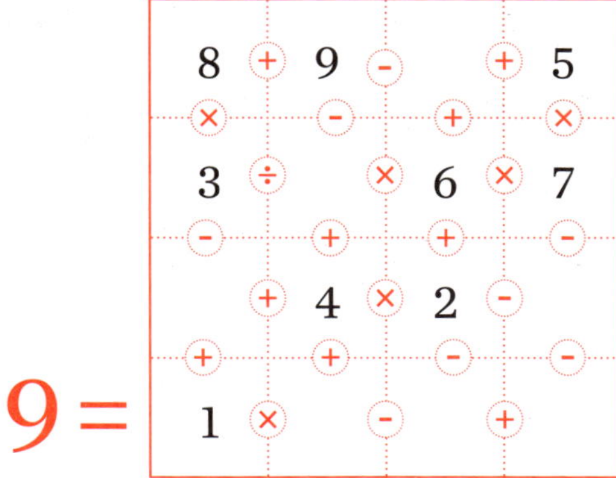

7	−		−	4	+	
+		+		+		−
1	−	8	÷	2	+	
×		+		−		+
	+	5	+	6	−	
−		−		+		÷
	−		+	9	+	3

= 9

The rules for these are easy. The answers, well, they're a little tougher!

For each 4 × 4 grid, enter the numbers 1 to 16 so that each row and column equals the target number. As always, order of operations matters.

Can you break the gridlock?

If you don't like defacing this book, go to adamspencer.com/resources for a template to fill in.

93

1000000000000
0000000000000
0000000000000
0000000000000
0000000000000
0000000000000
0000000000000
000

10^{93} is called one trigintillion (in America), and is sometimes called one quindecilliard (in Europe and Australia) ... 'this feels like it's gone on for a trigintillion hours'. Be careful saying that in Australia because you actually mean 10^{180} hours!

In fact 10^{93} is also called triacontahenillion in the ancient Greek-based naming system of massive numbers.

As to how often you'll get to use these words, well given that it's at least 10,000,000 times the number of fundamental particles in the entire universe ... not all that often.

93

Swig'n'swindle

Now, I like a quality piece of food or drink as much as the next person, but this is ridiculous.

On 10 February 1355, two students from Oxford University, Walter Spryngeheuse and Roger de Chesterfield, were not happy with the drinks served to them by John Croidon at his Swindlestock Tavern.

You might think the word 'Swindlestock' would be enough to alert you to quality concerns, but Wally and Rodge took their bevvies and threw them in John's face.

Did it end there? No. In the ensuing two-day riot between students and the townsfolk, 93 people died – 63 students and 30 locals.

Eventually the dispute was decided in favour of the university and a punishment was applied that lasted for centuries. Each year on 10 February, St Scholastica's Day, the mayor and members of the council had to march bareheaded through the town, go to church, and pay a fine of one penny for each of the 63 students who were killed. This punishment lasted 470 years until 1825 when the mayor of the day said, 'Sod that, I'm not apologising for something that happened back in 1355'. A reasonable position, if you ask me.

The town and the university finally let bygones be bygones when on 10 February 1955 the university gave the mayor an honorary degree and the council made the Vice-Chancellor an Honorary Freeman at a special ceremony ... 600 years after the stoush.

93

Major FLOPs

The fastest supercomputer in China is the brutally powerful number cruncher, Sunway TaihuLight. It was the reigning world champ for 2 years from June 2016 to June 2018.

Its grunt is measured as 93PFLOPS, which probably doesn't sound as impressive as it really is. Especially if you don't know much about PFLOPs.

Okay, let me try to unpack this one for you.

When doing computer calculations, the number 2 is considered an integer, but if it has a decimal place, like 2.45 or 2.9 or even 2.0, it is called a floating point number.

A FLOP or Floating Point Operation is pretty much any mathematical operation (+ – × ÷ √ and so on) that involves floating point numbers.

So working out the value of $(6.55 + 7.9 \times 0.0036)/4.09$ involves 3 floating point operations, or 3 FLOPs.

Well, our good friend the Sunway TaihuLight can crunch lots of floating point operations ... really quickly. 'How quickly?' I hear you whisper quietly, perhaps nervously.

Glad you asked.

93.01 petaFLOPS, where *peta-* is the prefix for 10^{15} or one thousand million million, and the PS in FLOPS (as opposed to FLOPs) stands for 'per second'.

Yep, ol' Sunny crunches through the math at 93,010,000,000,000,000 operations *per second* as it does everything from climate modelling to advanced manufacturing.

But things move pretty swiftly in computing ...

The top 6 supercomputers worldwide in terms of petaFLOPS (when I wrote this ... and I can almost guarantee it's changed if you're reading this more than five minutes later!) are (drumroll) ...

Summit or **OLCF-4** at Oak Ridge National Labs in Tennessee, USA. More about this veritable monster on the next page.

Sunway TaihuLight at 93.01PFLOPS. It sits in the National Supercomputing Center in Wuxi, China

China's National University of Defense Technology (NUDT) **Tianhe-2** (Milky Way-2) at 33.86

The Swiss **Piz Daint**, Cray XC50 system at 19.59 and Europe's most powerful

Japan's Agency for Marine-Earth Science and Technology operates **Gyoukou** at 19.14

Titan system installed at the Department of Energy's (DOE) Oak Ridge National Laboratory, which packs what seems like a measly 17.59 petaFLOPS

Stop press!

Just as we were going to print, a rumour started circulating that a supercomputer was in the offing that would shatter all known records and put the good old US of A back on the top of the PFLOPs POP charts.

And lo the word was true. Welcome Summit, the first supercomputer to exceed 100PFLOPs – and when I say 'exceed', Summit clocks in at ... wait for it ... 200PFLOPs.

Its architecture comprises 9216 POWER9 22-core CPUs and 27,648 NVIDIA Tesla V100 GPUs – which is great if you like hot and heavy computer architecture talk.

But how can you even picture that?

Well, Summit is the size of two tennis courts and calculates at the same speed as if everyone on Earth was using 26 million hand calculators at once ... one operation per second.

When generating data, the system can write to memory at a rate of 2.2 terabytes per second. That's like saving the entire latest season of *Stranger Things* in 0.002 seconds.

And, each minute, 15,000 litres of water flow through the system to keep it cool.

According to online references, Summit is used for 'scientific research'. Yeah ... that old one.

93

Drill bits (day 1 ...)

Ten players stood in a circle for their daily drill.

But this isn't just any old team. For my *Top 100*, I've assembled the hottest sporting team of all time. Any sport, any gender, any era — I don't care. As long as your guernsey ends in a 3!

Our superstars stand in a circle wearing their shirts numbered 3, 13, 23, 33, 43, 53, 63, 73, 83, 93.

Each day a different player is the drill leader, which by delightful coincidence, corresponds to the chapter number and 'front position'.

Your task, coach, is to work out the order they stood in the circle each day given the list of clues provided by these fun-loving sporty puzzlers.

What's that? Why not just ask them? Where's the fun in that?!

Things could get a little spicy because our first drill will be led by the legendary number 93 for the NBA's Sacramento Kings, Ron Artest.

You just don't know whether you'll get the elite defensive maestro Ron, or the Ron Artest who received the longest suspension for an on-court incident in NBA history for his role in the 2004 all-in brawl dubbed 'Malice at the Palace'. Or, for that matter, if Ron will insist everyone call him by his more recently adopted name, 'Metta World Peace'.

Anyway, this first drill should be a cracker! Over to you, Metta.

By some estimates, football — soccer — is played by a quarter of a billion people in over 200 countries, making it the world's most popular sport.

93

So, tell me ... in which order did the players stand around the circle?

Game on!

- #93 stood at the front of the circle.
- #93 stood next to #53.
- #53 stood closer to #43 than to #3.
- #13, #63, and #73 stood together in some order.
- There were 2 people between #3 and #33.
- #13, #33, and #73 stood together in some order.
- From #3's perspective, #93 was on the left side of the circle.
- There were 3 people between #33 and #93.
- #13, #23, and #33 stood together in some order.
- From #53's perspective, #93 was on the right side of the circle.

→

5	+	6	×	4
+	5	×	5	×
8	+	8	+	7
×	7	–	8	=
8	×	9	=	28

Snakes alive!

→

5	+	6	×	4
+	5	×	5	×
8	+	8	+	7
×	7	–	8	=
8	×	9	=	28

→

5	+	6	×	4
+	5	×	5	×
8	+	8	+	7
×	7	–	8	=
8	×	9	=	28

→

5	+	6	×	4
+	5	×	5	×
8	+	8	+	7
×	7	–	8	=
8	×	9	=	28

Welcome to the game of mathematical snake!

In the following grids, you can make paths from the top-left corner to the bottom-right by moving between adjacent squares. You can move up, down, left or right, but you can never visit the same square twice. And you guessed it, the paths must trace out the correct equation to reach the target number in the bottom right-hand corner.

For each grid, there are 3 different paths that all trace out correct equations. Order of operations applies! To be honest, even finding one correct answer for each grid will get you a gold star.

But why not see if you can find them all?

Alright, I'll start you off with this example above for the number 28, then ... we're off to the races.

Have a go, then check your answers against those on the left here. Answers for the grids on the opposite page can be found at the back of the book. Good luck!

→

5	×	9	×	1
×	2	×	3	−
7	−	7	×	7
×	1	+	4	=
7	×	3	=	92

→

7	×	8	×	7
−	2	×	2	−
1	×	9	×	2
+	6	−	5	=
1	+	6	=	92

→

9	×	1	×	2
−	4	+	9	+
9	−	9	+	1
×	7	×	3	=
5	−	9	=	92

Dude, don't snub my dodecahedron

A regular triangle

A shape is 'regular' if all of its sides are of equal length and all its angles are equal. Check out the sidebar to see what I mean.

Now, a solid is 'Platonic' if it is made up of the same regular face and has the same number of faces at each vertex ...

  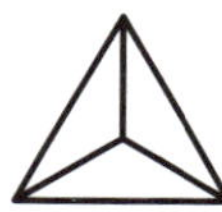  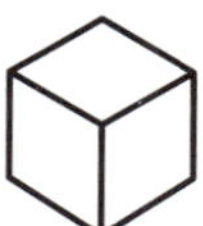

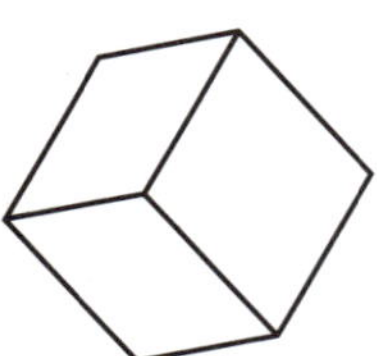

The faces of a cube are regular

Above we have an icosahedron, an octahedron, a tetrahedron, a dodecahedron and a cube. Ladies and gentlemen, I give you the 5 Platonic solids.

A solid is 'Archimedean' if it comprises faces that are all regular polygons (2-dimensional shapes) that meet at identical corners (vertices)*. There are 13 Archimedean solids, with some of the coolest names in mathematics like 'the small rhombicosidodecahedron' or 'the truncated tetrahedron'. Look them up online, they are totally lit (thanks to my daughter Olivia for providing this description which derives from 'legit' short for 'legitimate' meaning of great value)!

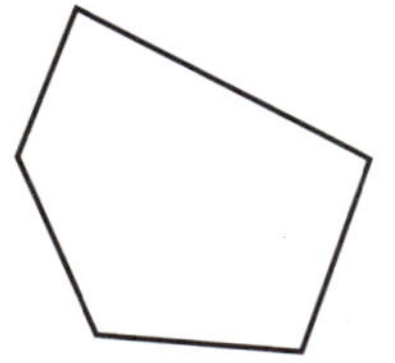

An irregular 5-sided shape

The Archimedean solid with the most faces, clocking 92 in all, is our old mate, the Snub Dodecahedron. You guessed it, that's the beast at the top of this page!

* Okay, the proper definition's a little more complicated than that, but you've got all the key points ...

92

Malaysia's Mahathir Mohamad is the world's oldest living head of state*

At least he was at the time this book was published. He was sworn in on 10 May 2018 at age 92. His historic victory in the 2018 Federal election marked the first time power had shifted from the opposition in 61 years. It also brought a smile to the faces of nonagenarians across the globe.

Queen Elizabeth II is not too far behind though, being just 285 days younger. The Tunisian Prime Minister Beji Caid Essebsi at 222 days younger than Queen Elizabeth is a close third.

* As of 2018. The oldest ever serving head of state however would be either Giovanni Paolo Lascaris, Grand Master of the Knights Hospitaller from 1636–1657; Abdul Momin, Sultan of Brunei from 1852–1885; or Enrico Dandolo, Doge of Venice from 1192 to 1205. They were all over 97 when they died in office.

91

Alphametic

Way back in my first book, *Adam Spencer's Big Book of Numbers,* I explored an awesome type of equation known as an 'alphametic'.

Alphametics, sometimes called cryptarithms, are sums written as words where the letters correspond to digits.

For example SEND + MORE = MONEY.

A couple of basic rules are that each digit is represented by only one letter and the first letter of any word can't have the value of 0, otherwise what looks like a 3-digit number would actually be a 2-digit number with a zero at the front.

If you write it out like this:

```
   SEND
+  MORE
-------
  MONEY
```

... and think about the rules of adding numbers, you will eventually work out the following:

O = 0, M = 1, Y = 2, E = 5, N = 6, D = 7, R = 8 and S = 9.

Substitute these numbers in and work out that our sum was:

```
   9567
+  1085
-------
  10652
```

Always looking to add value to your mathematical experience, it's time to meet 'Doubly-True Alphametics'. These are equations like:

```
  THREE
  THREE
    TWO
    TWO
+   ONE
-------
 ELEVEN
```

These can be solved as alphametics, and *also* hold up as sentences. Yes, 3 + 3 + 2 + 2 + 1 = 11.

You're probably still new to this game, so how about I give you some hints. In this alphametic, the value of THREE is 84611. There are also 9 letters used here, so of the digits 0 to 9, the one we do not use is the digit 5.

Substitute this information into the equation and see how you go.

To get you started, if you look down the right-hand column of the equation and substitute the possible values of O (it could be 0, 2, 3, 7 or 9; 5 is not used and the other values are already taken) you get possible values of N.

One by one, sub in the possible values of O and N and follow the logic through to see if you can get a solution ... or do you run out of possibilities? For example, if O = 7, the right-hand column gives you 1 + 1 + 7 + 7 + 1 = 17 so N = 7. This uses 7 twice, so we can exclude this.

If you manage to get an answer here, well done. If you can't yet, you'll be trying this alphametic *without* clues much later in the book and I'm backing you then!

As a general rule, alphametics take a while, but the logic isn't that difficult to follow.

You just need those very important mathematical skills of rigour and patience.

So to really help you work on your rigour and patience, I've decided to litter the rest of the book with doubly-true alphametics!

91

To get us started, solve:

```
      ONE
     FIVE
      TEN
   ELEVEN
 NINETEEN
+FORTYFIVE
---------
NINETYONE
```

Let me warn you now: the alphametics at 91, 81, 71 all the way down to 1 are probably the most infuriating and challenging puzzles in this entire book.

At the same time, this makes them without doubt the most impressive of all the nuts if you crack them.

As we count down, I'll give less and less of a hint and who knows, by the end of the book you may be solving them entirely hint free.

And I should give some props here.

In assembling these mind-benders I've relied on the work of three alphametic alpha-males: mostly Mike Keith who crunched *millions* of examples to compile the definitive list of all 266 DTAs that have a unique solution; Truman Collins who created an online alphametic solver through which I've verified my solutions and the godfather of DTAs; and Steven Kahan who, in 1969, found the very first doubly-true alphametic.

Good luck! But if you choose to pass these toughies by, on this one occasion I won't judge you.

Again, as a starting hint, I'll give you the value of NINETYONE — it's 828310483.

Substitute this into the alphametic and see if you can deduce the values of the remaining numbers.

Have fun!

Ninety-one is a nifty variation of the solitaire card game

But you'll probably have to rock a *real* deck of cards, old-school style if you want to play it, since not many computers include this geeky version.

If you'd like to have a crack, the rules are pretty straightforward.

Deal out 13 piles of 4 cards each.

You can move the top cards from one pile to another, without regard for their suit or rank. Your job is to move them around so the top cards total — you guessed it — 91.

Spot cards — from ace to 10 — are taken at face value, and the kings are valued at 13, the queens at 12 and the jacks at 11.

The easiest and most obvious 'win' is 91 = 1 + 2 + 3 + 4 + 5 + 6 + 7 + 8 + 9 + 10 + 11 + 12 + 13 , but there's no rule which says the cards all have to be different.

The Wiki entry for the game has one suggestion: 91 = 13 + 13 + 13 + 13 + 1 + 1 + 1 + 1 + 5 + 5 + 5 + 10 + 10 = 13 × 4 + 1 × 4 + 3 × 5 + 10 × 2 = 52 + 4 + 15 + 20 ... but there are plenty of other combinations out there. Go for it!

$$478^8 + 524^8 + 748^8 + 1088^8 + 1190^8 + 1324^8 = 1409^8$$

In the year 2000 a mathematician called Scott Chase, who wouldn't sound out of place as a billionaire bad guy in a Batman movie if you asked me, crunched out this awesome result, which expresses an 8th power as the sum of eight 8th powers.

Great Scott, Scott!

Blankety blanks (#90)

Head on back to number 100 if you need a reminder of how these puzzles work. Rules on the side, pens at the ready … aaaand go!

6 × ([ ] − [ ])×(([ ] − [ ])÷([ ] − [ ]) + [ ]) = 90

((8 − 5)×([ ] − [ ]) + [ ])×([ ] + [ ] + [ ]) = 90

([ ] − [ ])×([ ] + [ ])×(5 − ([ ] + [ ]) ÷ 6) = 90

(6 + 4)×(([ ] − [ ]) ÷ [ ] + [ ])×([ ] − [ ]) = 90

((([ ] − [ ]) × [ ] + [ ])×([ ] + [ ]) − [ ]) ÷ 2 = 90

Reach the target number by filling in the blanks in each equation.

A completed equation must incorporate each and every digit from 0 to 9, including the digits in the answer (so here you must place 1, 2, 3, 4, 5, 6, 7 and 8).

You cannot move the operations (+, −, ×, ÷) found between blanks.

Order of operations applies.

89

$8^2 + 9^2 = ...$

One awesome repository of online maths geekiness is the prodigious John D. Cook.

Apart from being a professional mathematician who does all sorts of maths consulting for clients like Amazon, Google and Microsoft, John absolutely bombards Twitter with maths facts via various handles.

It was one of his tweets about Number Theory via the handle @AlgebraFact that first alerted me to this cool bit of maths involving the number 89.

Take a whole number, any number, and add up the sum of the squares of its digits. When you get this sum, do the same to it again. And keep going and see what happens.

For example, the number 133 becomes $1^2 + 3^2 + 3^2 = 1 + 9 + 9 = 19$ and 19 becomes $1^2 + 9^2 = 1 + 81 = 82$.

You should be able to see that starting with 133 you get the numbers 133, 19, 82, 68, 100, 1, 1 ... which clearly stays at 1 forever more.

But the number 89 cycles through 89, 145, 42, 20, 4, 16, 37, 58, 89, 145 ... and clearly gets stuck in this loop.

In a mathematics article called 'A set of eight numbers' published back in 1945, the very brainy Arthur Porges showed that no matter what number you start with, you either descend to 1 or get caught in this 8 number loop.

How did he prove this? The method is really smart. It wouldn't be enough to test the procedure starting with 1, which clearly stays at 1, then testing 2 which goes 2, 4 and loops; then 3 which gives us 3, 9, 81, 65, 61, 37, then cycles and so on.

John D. Cook also maintains a very cool and very geeky blog over at his website: johndcook.com/blog. Grab your coffee and pencil and get your geek on!

While it might be encouraging if all of the numbers up to

1,000,000,000,000 did one of these two things, how do we know there isn't a counterexample out there somewhere?

So Porges first showed that for big numbers, like 1,000,000,000,000-sized, regardless of what value they actually take on through the chain, we can show that they keep getting smaller. In fact as long as the number has 4 or more digits, its next value must be smaller.

The biggest 4-digit number is 9999. Adding the squares of those digits gives $9^2 + 9^2 + 9^2 + 9^2 = 4 \times 9^2 = 4 \times 9 \times 9 < 10 \times 10 \times 10 = 1000$, so any 4-digit number becomes a number with 3 or fewer digits. This trick can be pretty straightforwardly applied for 5 or more digits too.

So we only have to consider all the numbers with 3 digits or fewer to convince ourselves we've looked at every case.

The 8 numbers in the cycle are the ones we've seen, but obviously if you landed on 73, this would generate 58, exactly like 37 did. So even though it might take another step to notice it, once you land on 73 or 85 or 98 and so on you are already trapped in the set of 8 numbers.

Crunching all possible starting numbers up to 10,000,000,000,000,000,000 gives us 8,816,770,037,940,618,762 numbers that get caught in the loop; about 88% of them.

The remaining numbers descend to 1.

Awesome.

Numbers that manage to reach 1 are called happy numbers; numbers that get stuck in the loop (and don't stop at 1) are called unhappy numbers.

Remember this: it's one of the definitions needed for our hexagon puzzles!

89

Sean's Syndesis

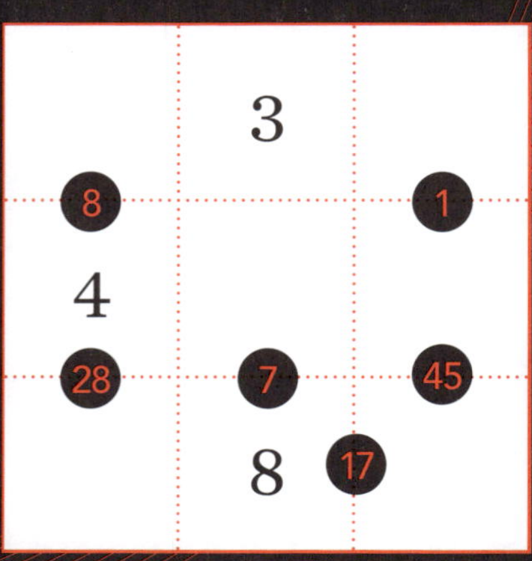

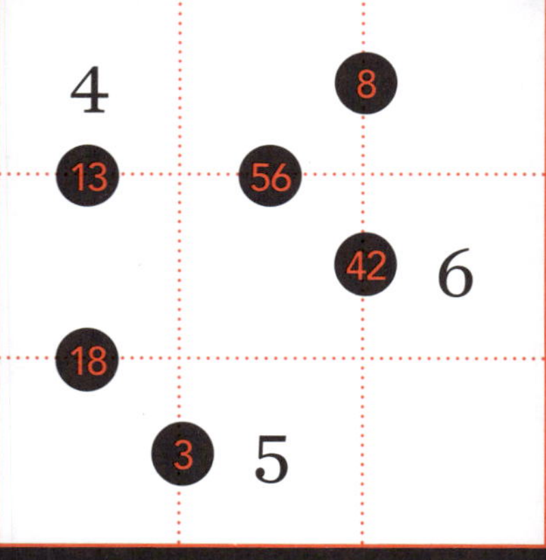

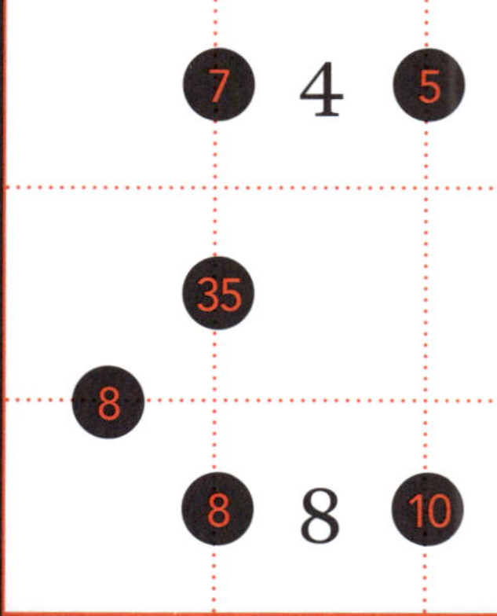

Enter the numbers 1 to 9 into each 3 × 3 grid so that each circled number is the result of adding, subtracting, multiplying, or dividing the two numbers in the cells it touches.

We've given you 8 clues here to get you started. At number 79, you'll get 7 clues and so on until you're crushing it all on your lonesome!

Head on back over to number 99 to read more about these awesome puzzles fresh from the mind of Sean Gardiner.

Gooaallllllllllllllll!

The movie *89* is billed as 'The true story of a sporting miracle, when Arsenal went to Anfield on the last day of the 1988/89 season needing to beat the best team in England by two clear goals to snatch the title'.

While the description might seem a little hyperbolic, it's pretty accurate. At kick-off for the final match in the 100th season of the Football League, Arsenal sat 3 points behind Liverpool but crucially also trailed them on goal difference*.

Arsenal (71 goals for, 36 against, goal difference 35) trailed Liverpool (65 for, 26 against, goal difference 39) by 3 points and 4 goals. But, crucially, every goal for Arsenal would be a goal against Liverpool. So the almost insurmountable 4 goal difference became effectively only 2 goals. Further, with more goals scored for, Arsenal, if level on points and goal difference, would claim the title on 'countback'.

Suffice to say Arsenal *really* needed to win by 2 goals or more. They had to do it at Anfield, where Liverpool had not conceded 2 goals in a game for over 50 matches.

And ... they did. With a goal to Michael Thomas deep into injury time with virtually the last kick of the match, Arsenal won 2-0 defeating Liverpool at home for the first time in 15 years.

This game was a postponed fixture out of respect for the 96 Liverpool fans who had died at the Hillsborough Stadium disaster and thus was played as a standalone title decider as opposed to the traditional final round when all matches were played at once. This only added to the surreal uniqueness of it all.

I must admit I don't really care for either team (go Man U!) so was comparatively unmoved at the time. But I can recall my good buddy and Liverpool devotee Simon Jones being absolutely speechless with shock.

Played, you Gunners.

* When teams are level on points, their 'goal difference' (goals scored across the season minus goals conceded) is what is used to separate them.

88

1. Rich Brian — 'Glow Like Dat' (54m)

2. Rich Brian, Keith Ape, & XXXTentacion — 'Gospel' (30m)

3. Rich Brian — 'Who That Be' (26m)

4. Joji — 'Will He' (24m)

5. Rich Brian — 'Chaos' (21m)

6. Keith Ape & Ski Mask the Slump God — 'Achoo!' (20m)

7. Joji — 'I Don't Wanna Waste My Time' (14m)

8. Diplo, Rich Brian, Young Thug, & Rich the Kid — 'Bankroll' (13m)

9. Rich Brian — 'Dat Stick' (Remix) (featuring Ghostface Killah & Pouya) (12m)

10. Higher Brothers & Famous Dex — 'Made In China' (11m)

Killer crossovers

If, like me, you're a massive fan of Asian-American crossover hip-hop acts like Rich Brian, Joji or Keith Ape, then you don't need me to tell you that 88rising is the name of an American media and music company.

If you're interested in 88rising's 10 most popular releases as at mid-2018, judged by YouTube hits, you're in luck. I've listed them at the side.

And if, like me, you're a massive fan of the internationally recognised constellations of the night sky, like Draco (the Dragon), Monoceros (the Unicorn) and Norma (my aunty – lovely lady ... no, actually the Carpenter's Levelling tool) then you don't need me to tell you that 88 is the number of constellations officially recognised by the International Astronomy Union.

As of mid-2018, and much less likely to change than a list of Asian-American crossover hip-hop artists ranked by YouTube views, the night sky's 10 biggest constellations are Hydra at 3.16% of the night sky, Virgo (3.14%), Ursa Major (3.10%), Cetus (2.99%), Hercules (2.97%), Eridanus (2.78%), Pegasus (2.72%), Draco (2.63%), Centaurus (2.57%), and last but not least ... Aquarius (2.38%).

Dancin' a jig

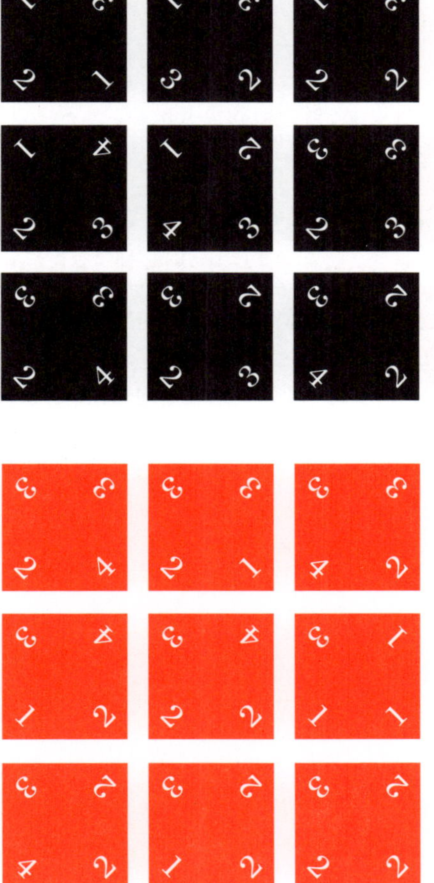

Arrange each set of 9 tiles into a 3 × 3 square so that adjacent tiles show the same numbers along their touching edges. Tiles may need to be rotated, but never reflected. Each set has only one solution. Check number 98 for more information.

For a downloadable cutout, head over to adamspencer.com.au/resources.

87

Number neurosis

According to the phobia wiki, the condition of *ogdokontaheptaphobia* is a fear of the number 87.

The word is supposed to come from ogdokonta, Greek for 'eighty', and hepta, Greek for 'seven'. Of course, it's not to be confused with 'hebdomekontahenophobia'.*

I don't want to sound insensitive, but I've never heard of anyone being nervous around the number 71. But it is a superstition in Australian cricket that batters should be wary when on 87. I recall reading in an excellent book once that, despite the superstition, 87 is actually a comparatively safe score to be on and that more batters have been dismissed for 85, 86, 88 and 89 than 87.

What's that?

You're wondering which excellent book I'm referring to?

Oh, that's right ... *Adam Spencer's Big Book of Numbers* by Adam Spencer.**

* The fear of 71.

** Apologies — even by my standards that was shameless!

New high score!

Welcome back to the game of High Scoring Equation where you fill in the blanks to win ~~real cash prizes~~ fame and glory! Have fun!

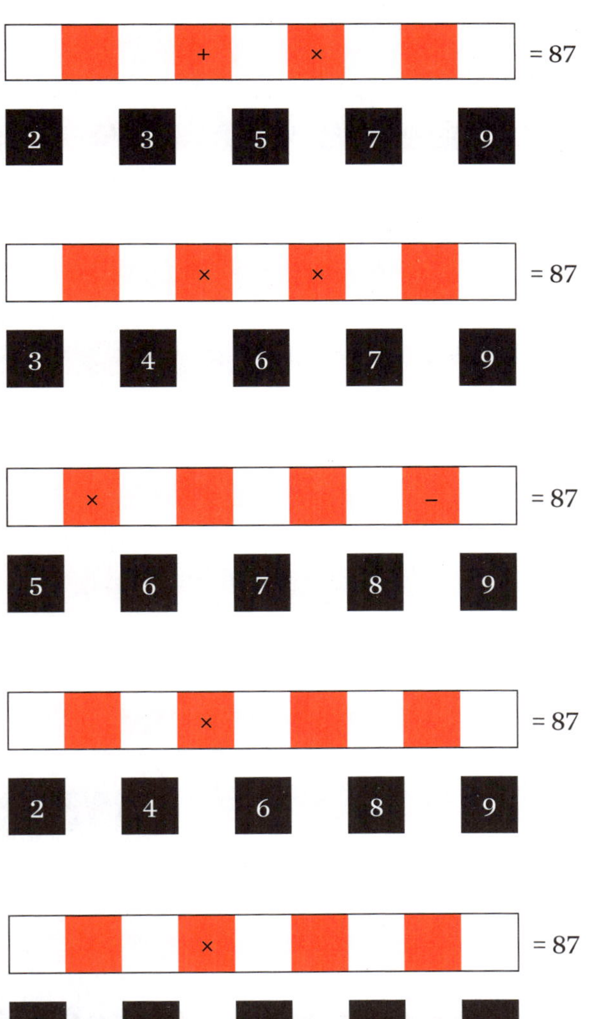

Reach the goal number by creating an equation using the provided numbers and your own choice of operators.

- The numbers should be placed in the white squares.
- Operations (+ – × ÷) are placed in orange squares.
- Order of operations matters, and you can't use brackets!
- Read the numbers off in order for your final score.
- Your goal is to find the equation that gives the highest score.

It's still early days, so I've given you 8 clues in total down the page.

Head back over to number 97 if you need a refresher on these, otherwise ... get cracking!

'86' is slang for refusing or getting rid of something ...

Why? Maybe because it's rhyming slang for 'nix'. But no one's really sure.

What we do know is that its first, verifiable use (according to no less a source than the Oxford English Dictionary) dates back to a 1944 book about 1920s movie star John Barrymore, who was infamous for his drinking: 'There was a bar in the Belasco building ... but Barrymore was known in that cubby as an "eighty-six". An "eighty-six", in the patois of western dispensers, means: "Don't serve him."'

The phrase gained more recent infamy when a member of US President Donald Trump's staff was refused service at a Virginia Red Hen restaurant as a protest against certain government policies.

The gesture, which polarised the American commentariat and made headlines around the world, was summarised on the staff notice board, alongside orders for stock and a note about the website, in the phrase '86 – Sarah Huckabee Sanders'.

More hexes!

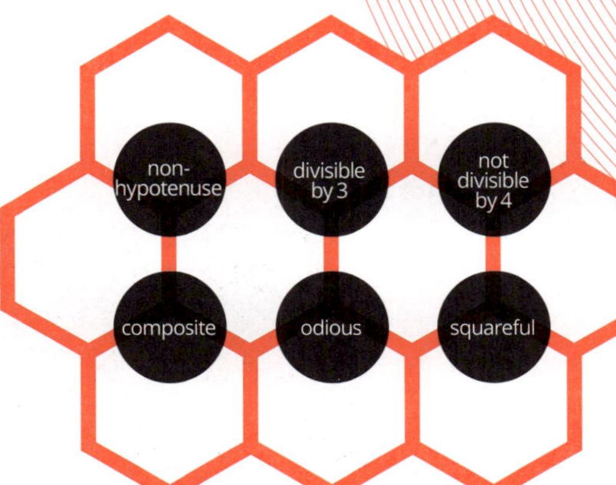

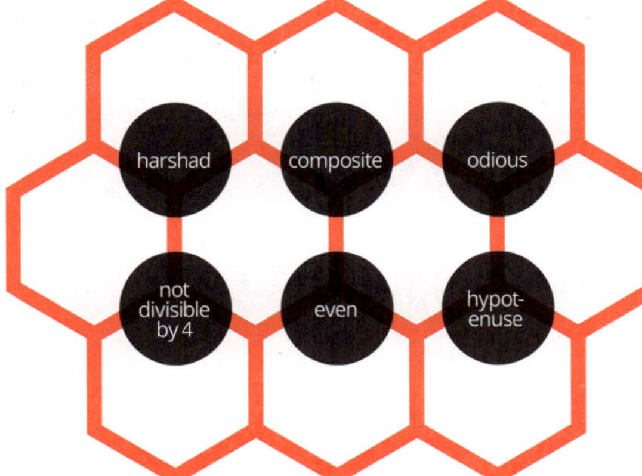

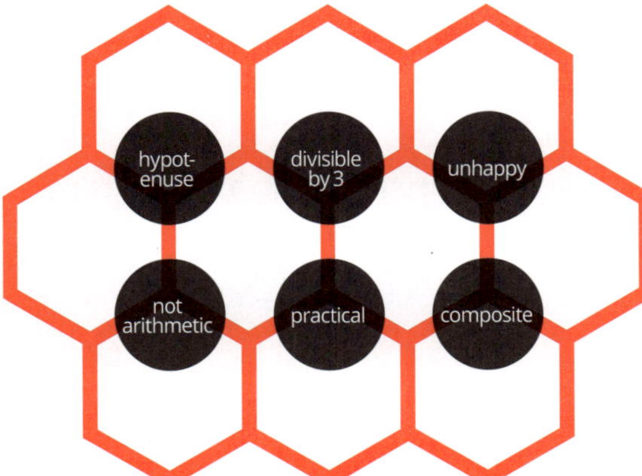

Place the numbers 81 to 90 into each hexagonal grid to satisfy all the rules.

Each cell is touching one or more categories in circles. The number in each cell must belong to all the categories touching it.

Need a hint? If the definitions are a bit tough for you, there's a full list of what these words mean at the end of the book.

And don't forget, if it's still challenging once you've looked up the definitions, these hexes are actually easier for smaller numbers — so maybe head to number 6 then work your way back through the hexes (16, 26, 36 and so on).

By the time you get back here, it should all make sense.

85

85% of Greenland is permanently covered in ice

Greenland is, according to friends of mine who have been, a gorgeous country. But it's not all that green. In fact, 85% of Greenland is permanently covered in ice.

They also have a pretty nifty, geometric flag which, just to add to the confusion, features no green whatsoever.

On the tiles

In each of these grids, finish tiling it so that it contains one of each pentomino type.

The 12 different pentominoes are shown here.

You may need to rotate and/or even flip some of them to get them to fit.

Head to adamspencer.com.au/resources if you'd like to download a copy of the puzzle to cut up!

84

COUNTEREXAMPLE TO EULER'S CONJECTURE ON SUMS OF LIKE POWERS

BY L. J. LANDER AND T. R. PARKIN

Communicated by J. D. Swift, June 27, 1966

A direct search on the CDC 6600 yielded

$$27^5 + 84^5 + 110^5 + 133^5 = 144^5$$

as the smallest instance in which four fifth powers sum to a fifth power. This is a counterexample to a conjecture by Euler [1] that at least n nth powers are required to sum to an nth power, $n > 2$.

REFERENCE

1. L. E. Dickson, *History of the theory of numbers*, Vol. 2, Chelsea, New York, 1952, p. 648.

Counting counter-conjectures

It might surprise you to learn that academic mathematics papers can be extremely long and remarkably hard to follow, even for trained mathematicians.

But at the opposite end of the spectrum we have the little beauty above.

84

In 1966, two mathematicians, Leon Lander and Thomas Parkin, announced a discovery involving the number 84.

The great Euler had suggested that while we have equations like $5^2 + 12^2 = 13^2$ or $3^3 + 4^3 + 5^3 = 6^3$ or, thanks to Scott Chase back at 90, the impossible-without-a-computer $90^8 + 223^8 + 478^8 + 524^8 + 748^8 + 1088^8 + 1190^8 + 1324^8 = 1409^8$ where the number of terms added together on the left-hand side is equal to the power that all the terms are raised to, you could never get a cube written as the sum of two cubes or, say, a sixth power written as the sum of four six powers.

Euler thought that in equations like these the number of terms added together would have to be at least the number of the powers involved.

Well, Lando and Parks (which I'm sure is *not* what their friends called them) shattered that thought, thanks to the help of a CDC 6600, which has been described as the first ever supercomputer. They calculated that $27^5 + 84^5 + 110^5 + 133^5 = 144^5$.

And thus the then-shortest paper in the history of mathematics was born.

In 1988, using a much gruntier computer than the good old CDC 6600, Roger Frye discovered that $95{,}800^4 + 217{,}519^4 + 414{,}560^4 = 422{,}481^4$. This is another example that disproved Euler's conjecture, but sorry, Rodge, Lando and Parks had you covered on that front by over 20 years.

As we count down to 1 in this *Top 100* we will feature a few more smash hits from the amazing back catalogue of what we call 'Diophantine equations'.

Stuck in gridlock?

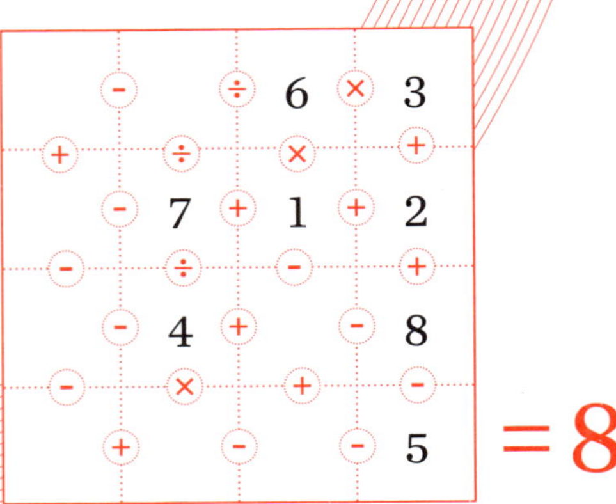

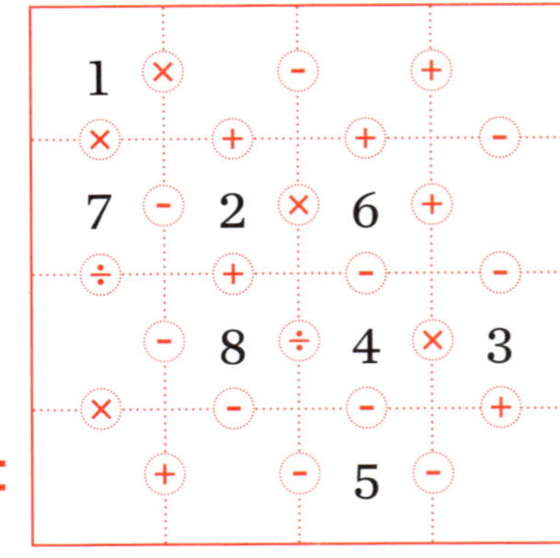

The rules for these are easy. The answers, well, they're a little tougher!

For each 4 × 4 grid, enter the numbers 1 to 16 so that each row and column equals the target number. As always, order of operations matters.

Can you break the gridlock?

If you don't like defacing this book, go to adamspencer.com/resources for a template to fill in.

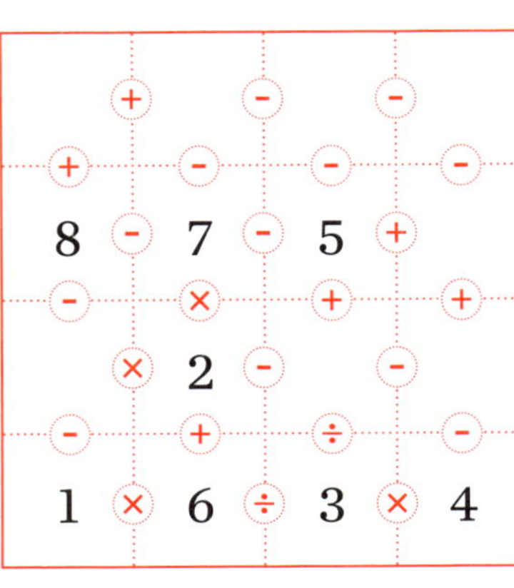

You know, you can be so 84-faced sometimes ...

I think it's fair to say that you're probably familiar with a cube.

A 3-dimensional cube can be thought of as six 2-dimensional squares joined together in the way kids do at school.

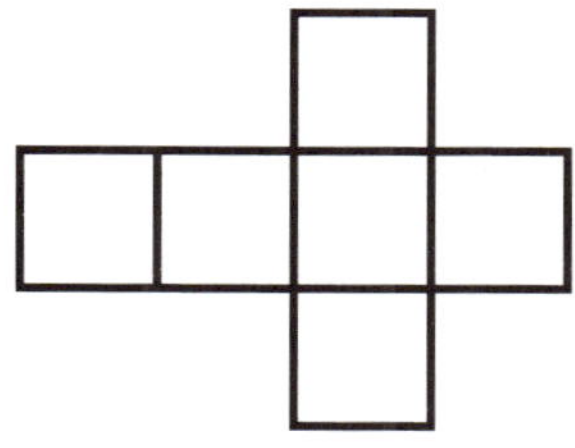

Simply fold the 6 squares in this 'net' to get ... a cube! (You know, like the thing at the side here ...)

The natural extension of a 2D square into 3 dimensions, the cube has 6 faces, 8 vertices (a fancy word for corners) and 12 edges.

Okay ... here we go ... if picturing things in higher dimensions makes you a little bit queasy, you might want to grab a glass of water.

While it is harder to picture, we can extend a 3D cube into 4 dimensions. It might be easier if we start with the 4D cube and unfold it, then fold it back up again ... trust me on this.

In the same way the 3D cube can be unfolded into the net we had above, a 4D cube could be unfolded into a collection of 3D cubes like the one on the bottom right, which is called a Dali cross (and yes, it is named after the artist Salvador Dali, but that's a whole other story ...)

You can then re-fold the 8 cubes into their 4-dimensional form and, as much as we can try to picture it on a 2D page with our 3D minds, you get something that ... well I can't really say 'something that looks like this' but perhaps 'something we can represent like this' ...

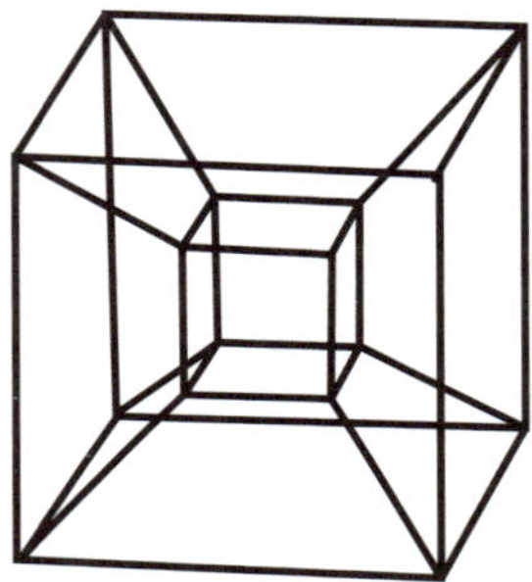

We call this a 'tesseract' or 4-cube. Built out of 8 cubes, it has 24 faces, 16 vertices and 32 edges, but it is also composed of 8 cells which are the 3-dimensional cubes from the net above.

Mathematically, there is nothing stopping you joining together a series of 4-dimensional tesseracts to get a 5-cube which exists in, you guessed it, 5 dimensions.

At this stage our 3D brains are hopelessly under-equipped to actually 'picture' this object, but mathematicians use diagrams to help capture some of the essential features.

Salvador Dali was fascinated by geometry and his 'passion for the fourth dimension' was famously immortalised with a massive hypercube serving as a crucifix to Christ in his *Corpus Hypercubus* (1954).

Get your geek on and check it out next time you're in New York at the Metropolitan Museum of Art.

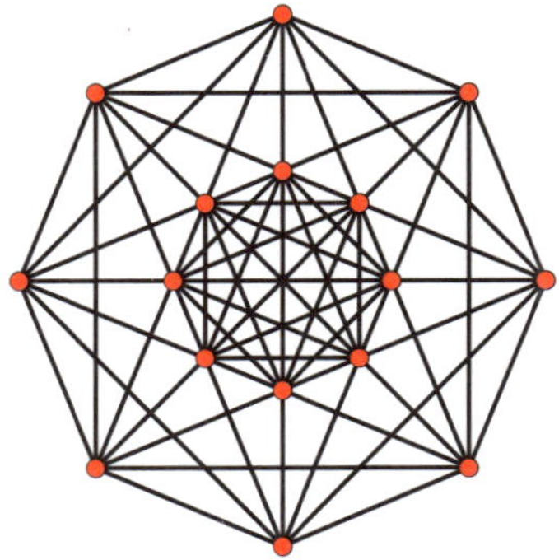

In the same way as the picture of a cube is what a 3-dimensional object looks like when projected onto a 2-dimensional page, this diagram is one of the possible projections of the 5-cube into 2 dimensions. Built out of 10 tesseracts, it has 80 faces, 80 edges and 32 vertices, but it also has 40 3-dimensional cells which are cubes and 10 4-dimensional faces which are tesseracts. (I'll be honest, it's not really helping me either!)

And on and on we can go; folding 12 5-cubes together to get a 6-cube and then, after another tall glass of water or perhaps something a little bit stronger, we can take 14 6-cubes and fold them up into a 7-dimensional object we call a 7-cube which, among other things, has 84 5-dimensional penteract faces.

Look at her ... isn't she gorgeous!

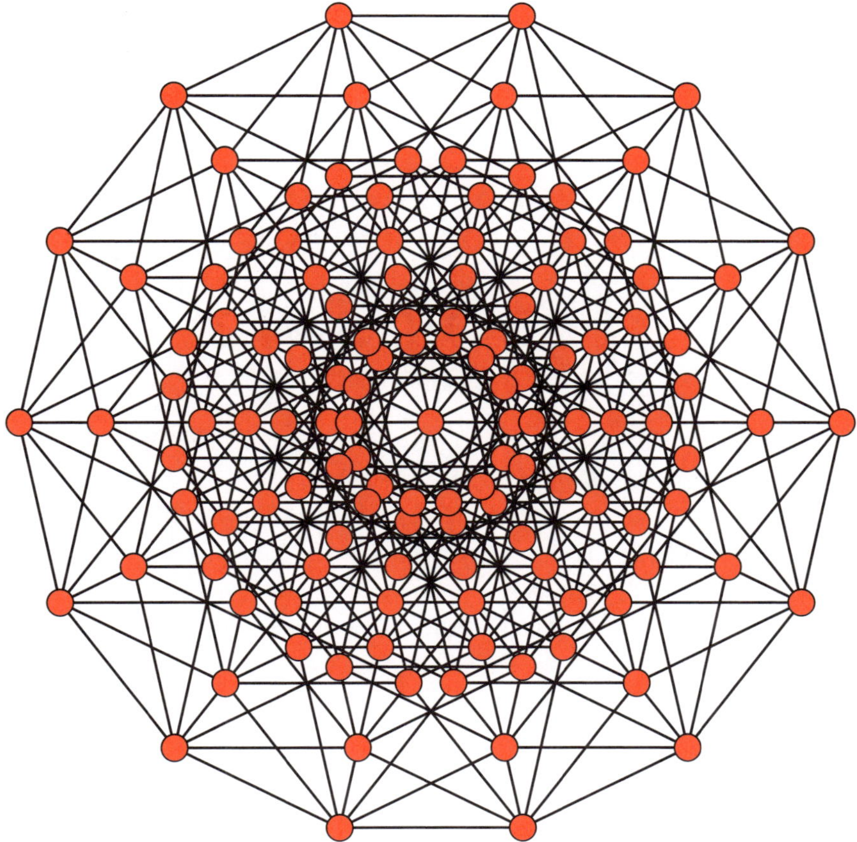

83

The big swim

A school swimming pool can hold 83 megalitres of water*.

Now, we know that a megalitre is 1 million litres. The pool can be filled by any or all of 3 pipes.

The most commonly used pipe is Pipe A which, by itself, takes 12 hours to fill the pool. The largest pipe, Pipe B, can fill the pool in 8 hours, while the emergency Pipe C would take a full 24 hours to do the job by itself.

If we opened all 3 pipes at once and they all flowed at the rate they flow when opened individually ... how long would it take to fill the 83 megalitre pool?

* Hey, it's a big school, okay? A standard Olympic swimming pool contains 2.5 megalitres of water.

Megalitres are nice and big, sure, but one *gigalitre* equals 1000 *million* litres.

Sydney Harbour? Glad you asked. It holds about 500 gigalitres.

Drill bits (day 2) ...

To ice hockey now, and today's drill will be led by Czech #83 Aleš Hemsky, who played a key role in the Edmonton Oilers charging into the 2005—06 Stanley Cup Finals, where they eventually lost in Game 7 to the Carolina Hurricanes.

In which order did the players stand around the circle?

- #83 stood at the front of the circle.
- #83, #3, and #33 stood together in some order.
- There was 1 person between #93 and #73.
- #93, #53, and #73 stood together in some order.
- From #73's perspective, #53 was on the right side of the circle.
- There were 3 people between #33 and #93.
- #93 and #13 stood the same distance from #43.
- #13, #43, and #63 stood together in some order.
- From #93's perspective, #3 was on the right side of the circle.

Ten players stood in a circle for their daily drill.

They are numbered 3, 13, 23, 33, 43, 53, 63, 73, 83 and 93.

Each day a different player is the drill leader, which by delightful coincidence, corresponds to the chapter number and 'front position'.

Your task, coach, is to work out the order they stood in the circle each day given the list of clues provided by these fun-loving sporty puzzlers.

Game on!

82

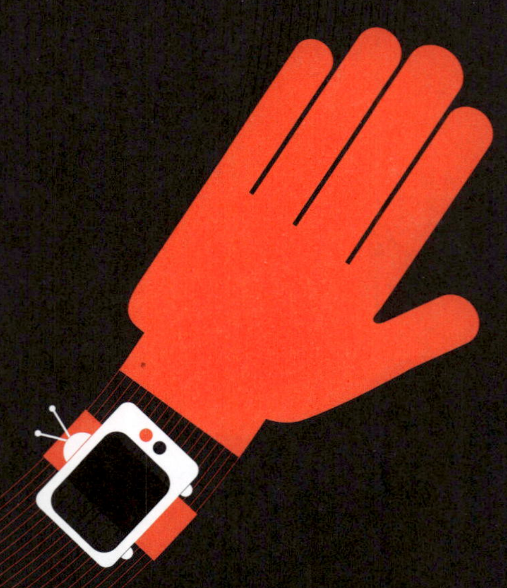

Introducing the Seiko TV watch ...

What could be more cutting-edge-80s than the Seiko TR02-01 Black and White TV Watch?

It was first available to the public in Tokyo in the summer of 1982 for a cool ¥108,000. At the time, that would have been roughly AUD$500, which is more than AUD$1730 today!

Snakes alive!

Welcome back to the game of mathematical snake! Your answer for each of these numbers is 82.

→					
→	9	–	4	+	2
	×	2	+	2	–
	4	×	4	×	5
	×	8	–	8	=
	7	–	7	=	82

→					
→	5	×	5	×	1
	+	8	+	8	+
	4	+	2	–	8
	–	2	+	7	=
	8	×	4	=	82

→					
→	2	–	5	×	7
	–	4	×	5	×
	2	+	6	×	3
	×	6	–	4	=
	7	+	8	=	82

In these grids, you can make paths from the top-left corner to the bottom-right by moving between adjacent squares.

You can move up, down, left or right, but you can never visit the same square twice.

And you guessed it, the paths must trace out the correct equation to reach the target number in the bottom right-hand corner.

For each grid, there are 3 different paths that all trace out correct equations.

In these very difficult puzzles, the order of operations applies.

Head on back to number 92 (if you haven't been there already) should you need a refresher.

81

Bloody good job

On Friday 11 May 2018, 81-year-old James Harrison donated blood for the last time.

He'd been doing it for over 60 years, averaging pretty much once a fortnight, for a total of 1173 donations.

By any measure that's a selfless thing to have done. But because of an incredibly rare combination, it's estimated James Harrison's blood did an astonishing amount of good.

Your blood is either 'Rh positive' or 'Rh negative' depending on whether your red blood cells have a protein called Rhesus Factor on their surface. Rh positive means you have the protein, which is the case for about 85% of people.

But if a pregnant woman with Rh negative is carrying a baby with Rh positive blood, her body misreads her baby's blood as something to be fought against and produces antibodies to destroy what it sees as an invader. This can have terrible effects on the unborn child.

In fact, before scientists discovered this condition (called HDN), thousands of babies died every year from this biological response. James Harrison's blood contains a crucial component needed to fight HDN and he produces it by the *truckload*! It is thought that *every dose ever created* in Australia to fight HDN has some of James Harrison in it.

As to why he is such a gigantic producer of this life-saving compound, some people think it might be because as a 14 year old he underwent major surgery and needed a massive transfusion of his own.

What we do know is that if you crunch the stats on the number of babies who could have caught HDN since the 1960s and how dangerous a disease it is, this one man has saved anything up to 2.4 *million* young lives.

As he donated for the 1173rd and last time, he admitted, 'It's a sad day for me. The end of a long run.'

Played, James Harrison.

The average 70 kilogram human contains approximately 5.5 litres of blood.

If you're a healthy geek aged 18—81 reading this book, consider visiting your local Red Cross and sharing some of the red stuff! You've got plenty to spare.

It takes 15 minutes to donate and 45 minutes for the appointment. You can donate every 12 weeks and *yes* you do get a biscuit!

Doubly-true (still!) alphametics

Back at 91 we met the alphametics.

Now, you're in one of two camps. Either you tackled the original alphametic in all its unadulterated glory and pain, and good on you. Or you said 'no thanks' to that and had a go at the easier version with the hint I gave you.

Obviously there is a third camp — those who didn't try either version — but I assume they turned the page about 60 words back when I first mentioned 'alphametics' again.

So here we go! At 81 on our top 100 countdown:

```
          O N E
        N I N E
    T W E N T Y
+     F I F T Y
---------------
    E I G H T Y
```

You might be wondering why this equation is our example for 81 when it doesn't contain the term 81. Well, it contains an EIGHTY and a ONE and when it comes to 'doubly-true alphametics', our options are rather limited.

If you're confident, please take this on in the raw. Good luck. But if you're still getting started ... check out the hint on the right.

I'll tell you that EIGHTY translates as 450,132. Substitute those letter values in and see how you go.

80

Weighty eighty

What do you notice about the list of numbers: 1, 2, 3, 4, 5, 6, 7, 8, 10, 11, 12, 13, 14, 15, 16, 17, 18, 20, 21, 22 ... 86, 87, 88, 100, 101, 102, 103, 104 ...

Yes, it's the list of all positive whole numbers that do not contain a 9.

The reciprocal of a number is when you divide one by that number. So the reciprocal of 4 is $1/4$.

If you add the reciprocal of all the positive counting numbers, $1/1 + 1/2 + 1/3 + 1/4 + 1/5 + \ldots$ the sum becomes infinitely large. This special sum is called the Harmonic Series and is one of the most famous sums in all of mathematics.

But here's something amazing. If you remove some of the terms from the Harmonic Series the sum adds up to a finite answer.

For example, the sum of the reciprocals of the list of 'nineless' numbers we have above is:

$$S = 1/1 + 1/2 + \ldots + 1/8 + 1/10 + \ldots + 1/18 + 1/20 + \ldots + 1/88 + 1/100 + 1/101 + \ldots$$

Grouping terms according to the number of digits in their denominator, we have:

$$S_n = \left(\frac{1}{1} + \ldots + \frac{1}{8}\right) + \left(\frac{1}{10} + \ldots + \frac{1}{18} + \frac{1}{20} + \ldots + \frac{1}{88}\right) + \left(\frac{1}{100} + \ldots + \frac{1}{888}\right) + \cdots + \left(\frac{1}{10^{n-1}} + \ldots + \frac{1}{8 \ldots 8\ [n \text{ eights}]}\right)$$

Now form another series, which is term-for-term greater than or equal to S_n, by setting each term with k digits in the denominator to $1/10^{k-1}$:

$$T_n = (1/1 + \ldots + 1/1) + (1/10 + \ldots + 1/10) + (1/100 + \ldots + 1/100) + \ldots + (1/10^{n-1} + \ldots + 1/10^{n-1})$$

Note that the number of k-digit numbers without a 9 is $8 \times 9^{k-1}$ (because there are 8 choices for the leading digit, and 9 choices for each of the other digits).

Hence:

$$T_n = 8 + \frac{(8 \times 9)}{10} + \frac{(8 \times 9^2)}{10^2} + \ldots + \frac{(8 \times 9^{n-1})}{10^{n-1}}$$

This is a geometric series — which some of you may remember from about Year 9 at school. Parents, if not, ask your 14 year old! This geometric series has first term 8, common ratio $9/10$, and n terms.

Therefore $T_n = 80[1 - (9/10)^n]$

As n approaches infinity,

$n \longrightarrow \infty$

T_n approaches 80.

$T_n \longrightarrow 80$

That is, the infinite series converges to 80. We can write this as $T_\infty = 80$.

But when we compare the terms of T_n to the terms of S_n, we see the terms in S_n are all equal or less than the matching term in T_n. So we know that:

$$S = (1/1 + \ldots + 1/8) + (1/10 + \ldots + 1/18 + 1/20 + \ldots + 1/88) + (1/100 + \ldots + 1/888) + \ldots < 80.$$

That is, S converges to a value less than 80.

Later in our *Top 100*, I'll reveal the actual value of S_∞ but for the moment, ain't that awesome!

On an unrelated note, the famous Rubik's Cube hit the markets in 1980 and in the ensuing 'Cube Craze' which lasted until '83, over 200 million cubes were sold. What a way to kick off a decade!

80

21, 22, 23, 24, 25, 26, 27, 28, 29, 30, 31, 32, 33, 34, 35, 36, 37, 38, 39, 40, 41, 42, 43, 44, 45, 46, 47, 48, 49, 50,

It takes about 40 seconds to count to 80

That's a very rough estimate, of course. Gee, thanks Adam. That's really useful.

Granted, you're not going to win pub trivia with this one, but this methodolgy used to calculate this number, from the numbermatics.com site, is interesting.

They calculate the number 'based on a speaking rate of half a second every third order of magnitude. If you speak quickly, you could probably say any randomly-chosen number between one and a thousand in around half a second. Very big numbers obviously take longer to say, so we add half a second for every extra ×1000'. Nifty.

They go on to note that they don't allow for involuntary pauses, loo breaks or 'the necessity of sleep' in their calculations.

Let's face it, if you need to add a sleep break while counting to 80, you should probably see a doctor.

Blankety blanks (#80)

Head on back to number 100 if you need a reminder of how these puzzles work. This time you'll get 8 clues in all. Rules on the right, pens at the ready ... hop to it!

(◯ + (◯ − ◯) ÷ (◯ − ◯) + ◯) × (7 + 1) = 80

((◯ + ◯) × (6 ÷ (◯ − ◯) + ◯) − ◯) × 5 = 80

(◯ − ◯) × (◯ + ◯) × (4 − (◯ + ◯) ÷ 6) = 80

((◯ + ◯) × 9 × (◯ − ◯) − ◯) ÷ (◯ + ◯) = 80

(◯ + ◯ ÷ (◯ + ◯)) × (6 × (◯ − ◯) − ◯) = 80

Reach the target number by filling in the blanks in each equation.

A completed equation must incorporate each and every digit from 0 to 8, including the digits in the answer (so here you must place 1, 2, 3, 4, 5, 6, 7 and 9).

You cannot move the operations (+, −, ×, ÷) found between blanks.

Order of operations applies.

79

The longest any two stick insects ...

have been recorded having sex is 79 ... seconds? Nope. Minutes? Nah. You're kidding? 79 *hours* ...?

Wrong again.

A pair of stick insects have been observed getting jiggy with it for ... a whole 79 days!

79

~~Seventy-eight~~ Seventy-nine

The human body is incredibly complex. But you'd expect that by 2018 we would have basically found all its parts. Well, think again.

In March 2018 doctors claimed to have found an entirely new organ. Not just that, if it is agreed to be an organ it could well be the largest organ in the entire body! Where was it hiding?

The 'interstitium' is the name being given to the network of interstitial fluid the lies beneath your skin. Interstitial means 'filling a very small gap' like a ray of sunshine peeking through a barely opened curtain. Scientists think we may have up to 10 litres of the stuff flowing around our bodies, and we only just realised this!

Before this discovery, most lists suggested there were 78 organs in the human body, but depending on your definition of organ, that number could be higher. That's why I can now confidently say there are, sort of, ish, 79 organs in your body.

79

What are the chances?!

Welcome to the town of Numberville.

A survey conducted showed that:

- 79% of the residents drink flat whites
- 51% drink lattes
- 90% drink water
- 80% drink tea

Now, no one drinks all four.

Tell me, what percentage of the residents drink coffee?

If you enjoy brainteasers such as these I highly recommend you point your browser to The Futility Closet (futilitycloset.com) which is packed with awesome things to stretch your grey cells.

Sean's Syndesis

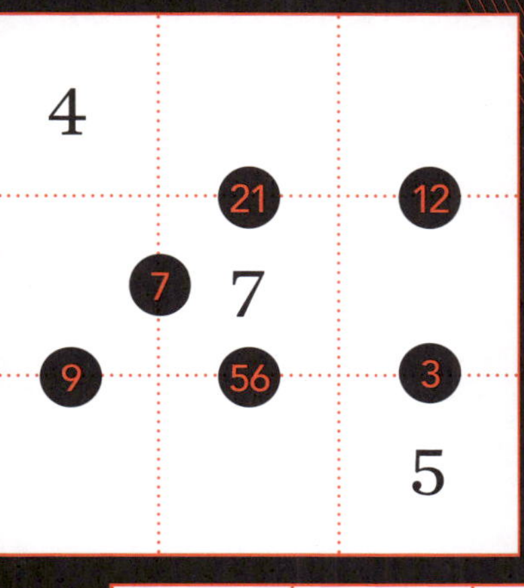

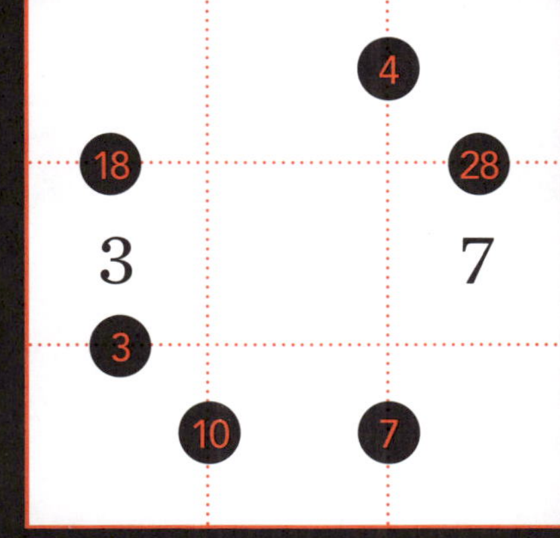

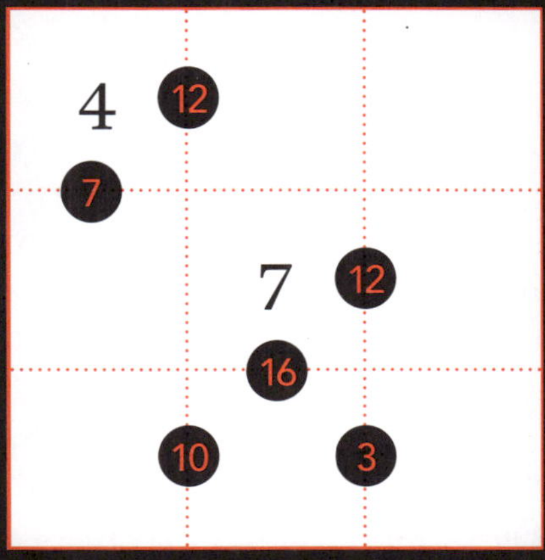

Enter the numbers 1 to 9 into each 3 × 3 grid so that each circled number is the result of adding, subtracting, multiplying, or dividing the two numbers in the cells it touches.

Head on back over to number 99 to read more about these awesome puzzles fresh from the mind of Sean Gardiner!

78

Good ol' 78s

There was a time when the vinyl record reigned as the supreme format of music.

Of course, many people still love vinyl and, unlike say, the rotary telephone, I'm pretty confident that most younger readers will be familiar with them (also, there's one on the cover of this book).

Although most 'normal' records after the early 1950s spun at a rate of either $33\frac{1}{3}$ or 45 revolutions per minute, back in the day – before my time, thank you very much – records were spun a fair bit faster, at 78 rpm.

Whatever rate they spin at, the principle's the same: a needle slots into a single groove which runs from the outside perimeter to the inside of the record.

Here's a question for you. Say I made a special record which, instead of a long, single groove, had two separate grooves – one that ran a single circle around the outer perimeter (the edge of the record farthest away from the centre), and one that did a lap of the inside perimeter (around the label). The inner groove is 6 cm from the centre of the record, the outer groove lies 18 cm from the centre.

If it takes 1.33 seconds for the needle to complete a loop of the inner groove, how long would it take the needle to complete a loop of the outer groove?

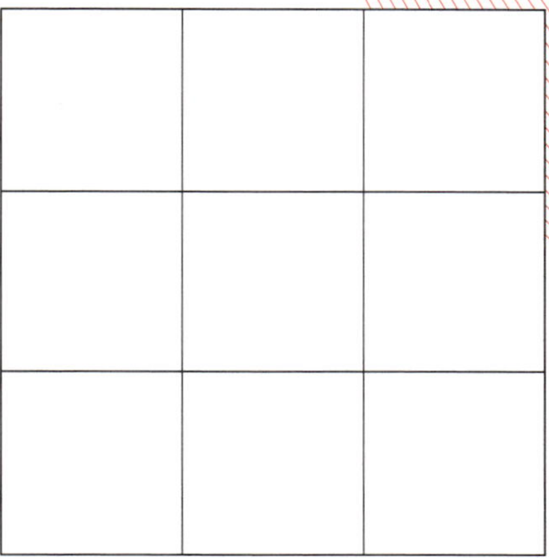

Prime-time magic

A square grid of numbers is called a 'magic square' if the numbers in each row, column and diagonal have the same sum. If you're not familiar with these beautiful objects, have a quick squiz online.

In his 1976 book *100 Numerical Games,* French puzzle maven Pierre Berloquin asks whether it's possible to construct a magic square using the first 9 prime numbers (counting 1 as prime ... which it strictly isn't ... but that's a whole other story, which you'll find at number 2 in our countdown).

These are: 1, 2, 3, 5, 7, 11, 13, 17 and 19.

Well, is it?

First, observe that the sum of these 9 numbers is 78. So if we create a magic square, each row, column and diagonal would have to sum to $78/3 = 26$. Get cracking!

You'll find the answer at the back of the book, of course.

Magic squares have been around for a long time.

They date back to at least 190 BCE in China, where they were primarily used for divination and astrology.

78

In 2018, Wally Scott-Smith finally retired from his role as the chief attendant of the Martin Place Cenotaph

The 96 year old had kept watch over the Sydney monument for 78 years.

He'd been keen to serve his country and enlisted with the Tank Corps in 1937, however a diagnosis of bowel cancer finished his military service before it had even started.

Still determined to do his part, he began cleaning the Cenotaph and, by 1946, he was asked to take over as chief attendant.

Although officially retired, he says, 'Next year I'll come, sit in a chair over there, and if the fellas aren't doing it right, I'll go crook.'

Good on you, Wally.

Tappin' a jig

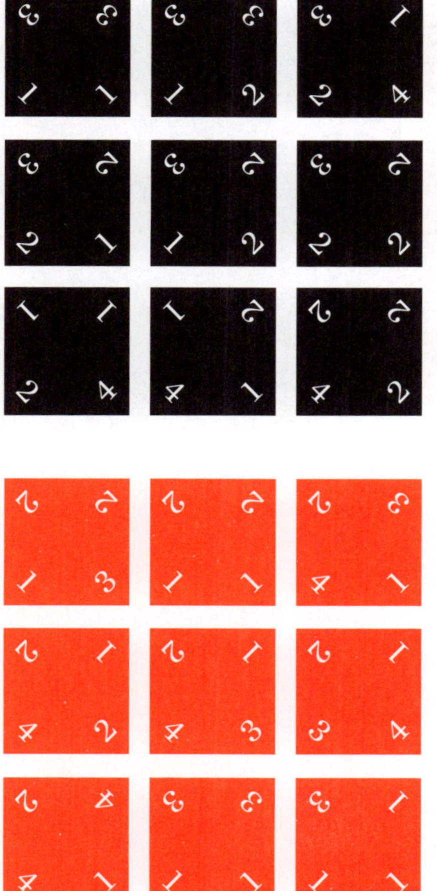

Arrange each set of 9 tiles into a 3 × 3 square so that adjacent tiles show the same numbers along their touching edges. Tiles may need to be rotated, but never reflected. Each set has only one solution. Check number 98 for more information.

For a downloadable cutout, head over to adamspencer.com.au/resources.

77

Australia holds the dubious distinction ...

of having 77 different species of marine animals threatened by entanglement and ingestion of plastic rubbish discarded into the sea.

Sadly, we are talking about 6 species of marine turtles, 12 species of cetaceans (whales, dolphins, porposies), at least 35 species of seabirds, dugongs, 6 species of pinnipeds (including seals), at least 10 species of sharks and rays, and at least 8 other sea-living species.

It's estimated that every square kilometre of ocean contains up to 18,000 pieces of plastic rubbish, but it's only when it washes up on our beaches or kills marine life that we notice the extent of the problem.

New high score!

Welcome back to the game of High Scoring Equation where you fill in the blanks to win ~~real cash prizes~~ fame and glory! Have fun!

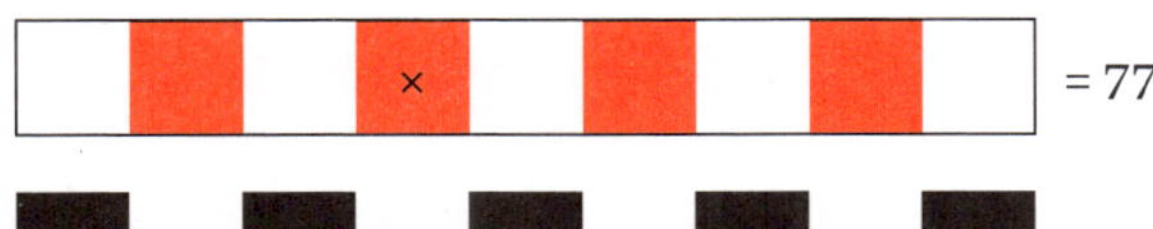

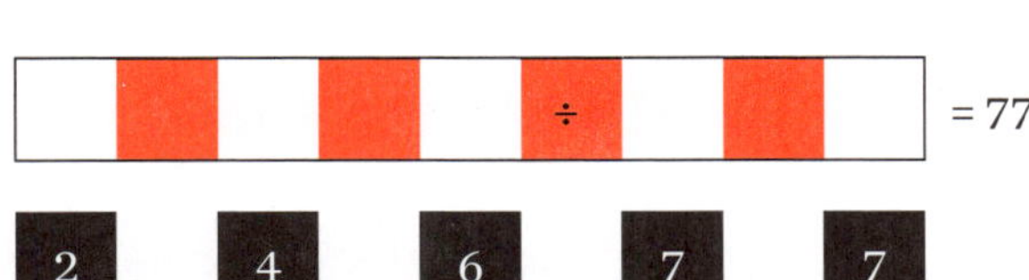

Reach the goal number by creating an equation using the provided numbers and your own choice of operators.

- The numbers should be placed in the white squares.
- Operations (+ – × ÷) are placed in orange squares.
- Order of operations matters, and you can't use brackets!
- Read the numbers off in order for your final score.
- Your goal is to find the equation that gives the highest score.

Head back over to number 97 if you need a refresher on these, otherwise ... get cracking!

77

Take me out to the geek fest

It's fair to say mathematicians look at the world in a different way.

On 8 April 1974 when Hank Aaron, legendary baseball hitter for the Atlanta Braves, smoked one over the left centrefield wall off the pitching of the LA Dodgers' Al Dowling, the max-capacity crowd of 53,775 at Atlanta-Fulton County Stadium went absolutely bonkers.

Aaron had just hit his 715th career home run and in doing so passed the record set by the legendary Babe Ruth who on 25 May 1935, at Forbes Field in Pittsburgh, Pennsylvania, hit his 714th and last home run.

Spectators hopped the fence and congratulated Aaron as he lapped the bases, fireworks erupted overhead, opponents shook his hand, a smile spread across his normally implacable face as an incredible weight lifted off his shoulders ... and at home plate, he was greeted by his mum!

But mathematician Carl Pomerance and a friend noticed something else; while sports fans saw 714 and 715 as the record number of home runs, these maths geeks noticed that the sum of the prime factors of 714 and 715 was the same.

A special class of numbers was born: Ruth-Aaron pairs — consecutive numbers whose prime factors have the same sum.

I know, it sounds very nerdy, but what does it mean?

Well, the first Ruth-Aaron pair is (5,6) because the prime factors of 5, being just 5, and 6, being 2 and 3, have the same sum: $5 = 2 + 3$.

For 714 and 715 it's a bit nastier but we have $714 = 2 \times 3 \times 7 \times 17$ and $2 + 3 + 7 + 17 = 29$.

Here's a fun one. Find the prime factors of 715 and show it forms a Ruth-Aaron pair with 714. While you're at it, show that 77 is involved in a Ruth-Aaron pair.

If you're feeling brave, grab a pen and paper and settle in for about half an hour of real fun (well, my sort of fun anyway).

The first Ruth-Aaron triplet is the numbers 89,460,924 89,460,925 and 89,460,926. Factorise these brutes and show they are indeed such a triple. I'll give you the hints that 89,460,294 has 5 prime factors under 30 and one massive one in the 8000s; 89,460,296 has 5 prime factors under 50 and one in the 8000s and 89,460,295 has only one (obvious) small factor but the other two are almost equal — I'll let you consult a list of primes to help you here.

Off you go! See if you can slug an arithmetic homer of your own.

As an addendum, Hank Aaron was one of the most dignified and humble champions in the history of any sport.

Take a moment to watch him interviewed online.

And definitely check out that 715th homer. He toasts it!

Tune in

Approximately 76.3 million people watched the final episode of the comedy smash *Seinfeld* on American television on 14 May 1998.

Impressive? Sure. But even those amazing numbers see the *Seinfeld* finale ranked only 3rd in American TV viewing records.

The final episode of *Cheers*, which aired on 20 May 1993, hit a whopping 84.4 million viewers.

But 'Goodbye, Farewell and Amen', the final episode of *M*A*S*H* that aired on 28 February 1983, pulled in an incredible 105.9 million sets of eyes.

At the time, the population of America was 234 million suggesting 45% of all Americans watched that show!

*M*A*S*H*, which aired on CBS for 11 years from 1972, was also numerically impressive for the fact that the TV series lasted significantly longer than the conflict on which it was based (that would be the Korean War, which lasted three years, one month and two days).

76

The NBA basketball team...

the Philadelphia 76ers were named by fan Walter Stalberg in a competition in 1963. He was inspired by the signing of the United States Declaration of Independence in that city in 1776.

Stephen Hawking passed away at the age of 76

He died on National Pi Day (in the US) — 14 March 2018, or 3.14.18.

It was also none other than Albert Einstein's birthday which, aside from being a cool piece of trivia, is what mathematicians call ... a coincidence.

More hexes!

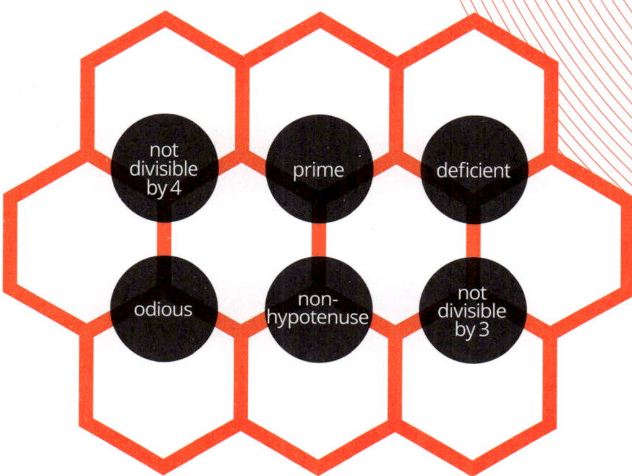

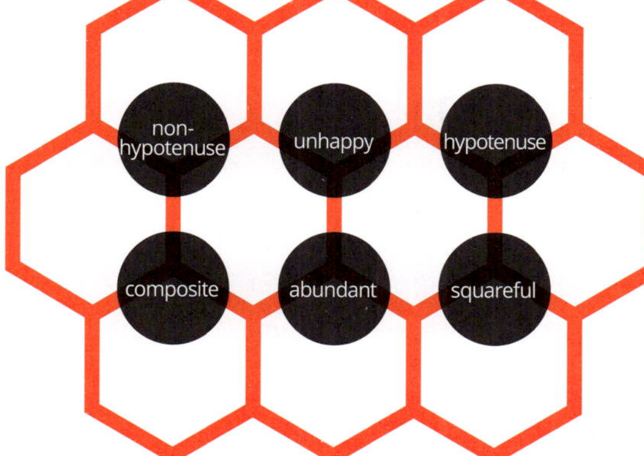

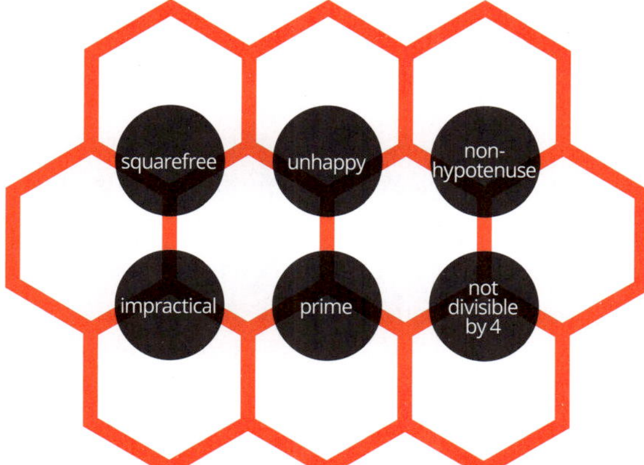

Now, don't forget that as we count down to 1, this class of puzzle actually get easier, so if you're struggling, skip ahead to 6 then work your way back through 16, 26, 36 and so on.

Place the numbers 71 to 80 into each hexagonal grid to satisfy all the rules.

Each cell is touching one or more categories in circles. The number in each cell must belong to all the categories touching it.

Need a hint? If the definitions are a bit tough for you, there's a full list of what these words mean at the end of the book.

75

Rock on!

The Pet Rock went on sale in December 1975.

That's right, folks, it wasn't until advertising exec Gary Dahl rocked up with a stone, some googly eyes and a hot glue gun that this cheap, clean and extremely boring paperweight of a pet came along.

They originally sold for $4 and were tragically discontinued in February 1976 due to falling sales ... presumably as people domesticated their own wild rocks by sticking googly eyes on them for next to nothing.

But don't feel too bad for Gary. In the 6 months they were on sale, he sold in excess of 1.5 million units carrying the original $4 price tag.

Gary wasn't the only entrepreneur on the block in 1975. Earlier that year, in Albuquerque, New Mexico, Paul Allen and Bill Gates decided to set up shop selling BASIC interpreters for the Altair 8800 microcomputer. They named their company Micro-Soft, and the rest, as they say ...

The Altair 8800 base model came in two versions: assembled and unassembled as a kit.

The kit price was US$439 (roughly $2000 nowadays) and US$621 if you didn't want to have to heat up your soldering iron.

On the tiles

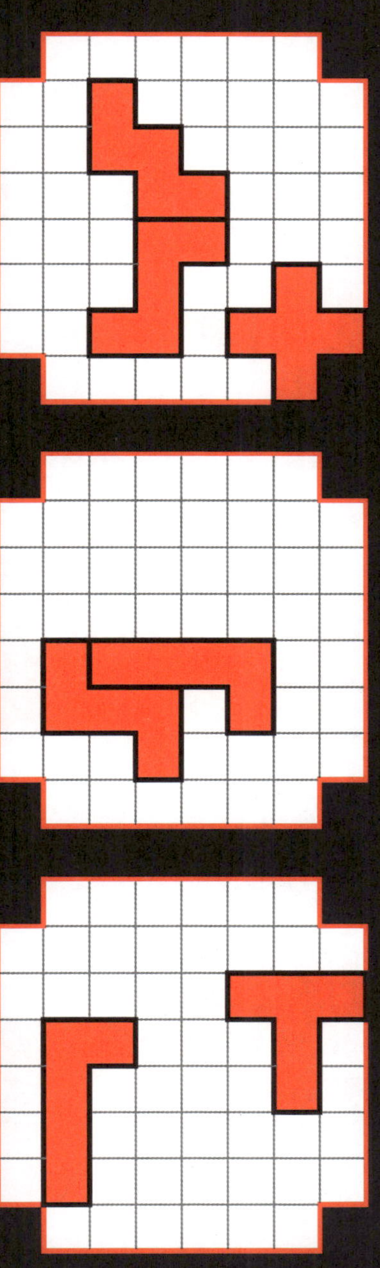

In each of these grids, finish tiling it so that it contains one of each pentomino type.

The 12 different pentominoes are shown here.

You may need to rotate and/or even flip some of them to get them to fit.

Head to adamspencer.com.au/resources if you'd like to download a copy of the puzzle to cut up!

Sunday, 27 August 2017 is a day Aussie Scrabble-lover David Eldar will never forget

He beat English champ Harshan Lamabadusuriya at Old Museum, Nottingham, 3-0 in a best of five final to win the world championship.

His best word in the final game, CARRELS, a cubicle desk used in schools, scored him 74 points. A good effort, but a long way short of the most points ever earned for a single word in a game of Scrabble – I'm referring, of course, to that famous moment when the legendary Karl Khoshnaw dropped CAZIQUES (black and red American tropical songbirds ... but I didn't need to remind you of that) earning a momentous ... 392 points!

Seventy-four no more

We'll talk a bit in this book about expressing a number as the sum of powers.

Trust me, you'll know a lot about this sort of stuff once we're done. And I'm totally cool with that.

Most of the examples you will see involve only positive whole numbers, or integers. But once we allow negative numbers as well, things get even more interesting. 'What, Adam? How could they get *even more* interesting?' I hear you cry. Well, fellow traveller, hold on.

Take the number 29. Using just positive integers we can write $29 = 3^3 + 1^3 + 1^3$.

But allowing negative numbers we can also consider $4^3 + (-3)^3 + (-2)^3 = 4^3 - 3^3 - 2^3 = 64 - 27 - 8 = 29$.

Note that adding a negative number cubed is the same as subtracting the cube of the positive version of the number.

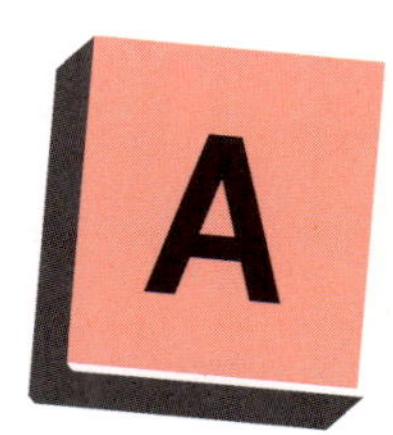

As we work through the list of all positive integers and ask if they can be expressed as the sum of 3 cubes we first notice that if the number n can be written in the form $n = 9k \pm 4$, where k is another whole number, then n can't be written as the sum of 3 cubes.

The proof of this isn't that hard to follow as long as you understand 'modulo arithmetic' which I'm afraid I don't have space in this book to explain*. If you're feeling adventurous and really can't wait a year to know why, then look the $n = 9k \pm 4$ proof up online and I think you'll like it.

Anyway that means the numbers 4, 5, 13, 14, 22, 23, 31, 32, 40,41, 49, 50 ... can't be expressed this way.

Mathematicians being the thorough types we are, who love a challenge, have gone through all the other whole numbers and tried to express them as the sum of 3 cubes.

* Pinkie promise it will feature in next year's tome ... #spoileralert!

74

Some of the solutions are easy, like 29 above. Some numbers have an infinite number of solutions: for example for every value of t you wish to use $(9t^4)^3 + (3t - 9t^4)^3 + (1 - 9t^3)^3 = 1$. Try this for a few values of t if you are so minded. It's really interesting.

Some require massive values like $30 = (-283059965)^3 + (-2218888517)^3 + (2220422932)^3$ which wasn't discovered until 1999 despite this having been a maths thing for a long time by then.

In fact it may well be that for every integer there is an infinite number of solutions, just spread out over a long distance and involving for the most part, gigantic values to be cubed. At the moment we just don't know.

But one by one they fell and until recently there were only 3 small numbers that hadn't succumbed to being expressed as the sum of 3 cubes; namely 33, 42 and 74.

The superb online video series 'numberphile' did an episode of the cubes problem and, inspired by that, Sander G. Huisman grabbed some serious computing grunt and using effectively 12.5 years of CPU processing managed to knock one of the outstanding suspects off the list.

Turns out that:

$$74 = (-284650292555885)^3 + 66229832190556^3 + 283450105697727^3$$

... all along. We just never realised it. More fool we.

Numbers 33 and 42 though? They're still out there laughing at us.

Unlock the gridlock

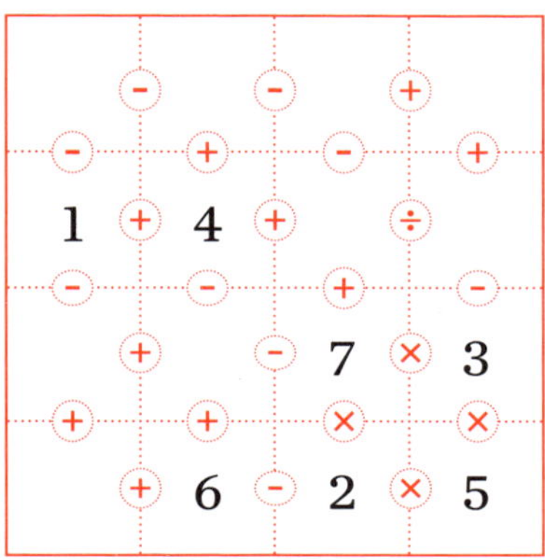

= 7

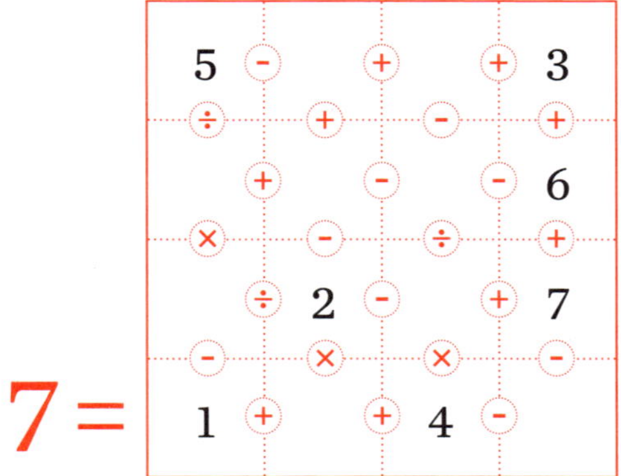

	−		÷	6	×	3
+		−		+		+
	−		+		×	1
÷		+		−		−
2	+	7	+		−	
−		+		−		+
	÷	4	÷	5	×	

= 7

The rules for these are easy. The answers, well, they're a little tougher!

For each 4 × 4 grid, enter the numbers 1 to 16 so that each row and column equals the target number. As always, order of operations matters.

Can you break the gridlock?

I'll start you off with numbers 1–7.

If you don't like defacing this book, go to adamspencer.com/resources for a template to fill in.

73

Spaaaaaaaaaaaace

In 2018 the value of the Hubble constant, the speed at which the universe is expanding, was announced as 73 kilometres per second per megaparsec[1].

The researchers who estimated the new speed used the Hubble space telescope to measure a series of stars every 6 months for 4 years. The class of stars, Cepheids, were between 6000 and 12,000 light-years away and were chosen because of the predictable way their brightness changes over time.

Interestingly, this result is different to another experiment that looked at the left-over glow from the big bang, the cosmic microwave background radiation, and came up with a range between 67 and 69.

While this may not seem much of a difference if you were, say, driving a car at 68 km/h versus 73 km/h, when it comes to understanding such a fundamental measure that drives our universe, small differences can mean very big things.

Watch this (rapidly expanding, possibly infinite) space!

[1] A megaparsec, of course, is equal to about 3.3 million light-years.

Mic drop

$$271^3 + 2^3 \times 3^5 \times 73^3 = 919^3 = 776{,}151{,}559$$

This is cool within itself, but is also an important observation when trying to prove a notoriously difficult suggestion in mathematics called Beal's conjecture.

Unfortunately I don't have time to go into Beal's conjecture here (next book! Promise!), but next time you're at a party and someone brings the subject up, feel free to throw in the gem above.

As their jaws drop, round off with a casual ... 'and of course, $3^4 \times 29^3 \times 89^3 + 7^3 \times 11^3 \times 167^3 = 2^7 \times 5^4 \times 353^3 = 3{,}518{,}958{,}160{,}000$.'

Then mic drop[2] and leave. You know the drill.

[2] People who monitor such matters say that the very first recorded 'mic drop' was by Judy Garland on a 1965 epsiode of *The Ed Sullivan Show*. You can find it on YouTube and yes, it's a total baller move.

Drill bits (day 3) ...

Our superstar, cross-sport, intercontinental mega team will today be led by American Football legend (and possibly the greatest guard in the history of the NFL), #73 John Hannah. Not to be confused with the Scottish actor of the same name from *Four Weddings and a Funeral*.

In which order did the players stand around the circle?

- #73 stood at the front of the circle.
- #93, #53, and #63 stood together in some order.
- #83, #13, and #23 stood together in some order.
- From #23's perspective, #33 was opposite them.
- There were 4 people between #73 and #3.
- #53 and #83 stood the same distance from #33.
- #93, #3, and #53 stood together in some order.
- From #33's perspective, #43 was on the left side of the circle.

Ten players stood in a circle for their daily drill.

They are numbered 3, 13, 23, 33, 43, 53, 63, 73, 83 and 93.

Each day a different player is the drill leader, which by delightful coincidence, corresponds to the chapter number and 'front position'.

Your task, coach, is to work out the order they stood in the circle each day given the list of clues provided by these fun-loving sporty puzzlers.

Game on!

72

How about a nice game of chess?

December 2017 saw one of the most spectacular displays of machine learning ever.

Google DeepMind's learning algorithm Alpha Zero was given the rules of chess.

Just the rules. No tactics or advice nor any famous past games to analyse.

Not preprogrammed with the millions of strongest opening combinations of moves that all decent chess computers carry as a matter of course.

I repeat, it was simply told the rules.

Starting from the state of 'Tabula Rasa' or 'a blank page', Alpha Zero then played games of chess against itself, learning from the results. Not just a few games, or a few hundred games. Alpha Zero played itself 4 million times in just 4 hours, playing thousands of games incredibly quickly and in parallel on 5000 of Google's ultra fast and powerful Tensor Processing Units or TPUs.

Using 'reinforcement learning', Alpha Zero kept track of good moves and positions that led to wins. Moves that were bad were given a very negative weighting to make sure they were not repeated.

Alpha Zero played a version of Stockfish, one of the world's strongest chess engines, used by sites like chess.com, in a 100-game showdown.

The good news for Stockfish was it managed 72 draws. The bad news for Stockfish was that Alpha Zero won all 28 of the other games!

And it wasn't the boring, brute force chess that computers often play when they grind a human into submission.

Alpha Zero, in just four hours, taught itself to play a style of chess that understood every opening and strategy humans had ever developed, and much higher, more beautiful ideas. Brilliant sacrifices no human would ever make, deeply complex tactical manoeuvres our minds had never imagined and our eyes had never seen in the thousands of years we've been playing this beautiful game.

As Peter Nielsen, who coaches World Champion Magnus Carlsen, said, 'The aliens came and showed us how to play chess'.

Game on.

In 2014, the internet's most visited chess site, chess.com, announced that over a billion live games had been played there, including 100 million correspondence games.

They calculated an average of 4.18 games were started every second around the clock since the site's launch in 2007.

I'm proud to say I'm a member. Yes, chess.com rocks!

72

Simply Eiffel

Although today it is regarded as an architectural and engineering masterpiece, at the time it was being proposed, the Eiffel Tower was very controversial.

One group of 'Artists against the Eiffel Tower' notably complained 'of this useless and monstrous Eiffel Tower ... imagine for a moment a giddy, ridiculous tower dominating Paris like a gigantic black smokestack, crushing under its barbaric bulk Notre Dame, the Tour Saint-Jacques, the Louvre, the Dome of Les Invalides, the Arc de Triomphe, all of our humiliated monuments will disappear in this ghastly dream.'

Dudes, don't hold back.

Well, the tower went ahead and the rest is history ... and tourism ... and all sorts of stuff.

As his own little statement on the matter, and to jam one in the eye of the artists, engineers, architects and general public who had opposed him, Gustave Eiffel engraved the names of 72 great French scientists, engineers and mathematicians on the sides of the tower — just under the first balcony, in letters about 60 cm high which were painted in gold.

An astonishing 7 million people visit this iconic 'giddy, ridiculous tower' each year.

In fact, an estimated quarter of a billion people have said 'bonjour' to the Eiffel Tower since its opening in 1889.

Snakes alive!

→				
7	–	4	+	3
+	2	×	2	–
7	–	5	+	3
×	4	×	8	=
9	×	8	=	72

→				
4	×	4	×	4
+	3	–	3	×
4	+	6	×	9
–	4	–	2	=
3	–	4	=	72

→				
1	+	7	+	2
×	1	×	4	+
9	+	4	+	7
–	8	–	7	=
9	+	4	=	72

In these grids, you can make paths from the top-left corner to the bottom-right, by moving between adjacent squares.

You can move up, down, left or right, but you can never visit the same square twice.

And, you guessed it, the paths must trace out the correct equation to reach the target number in the bottom right-hand corner.

For each grid, there are 3 different paths that all trace out correct equations.

And don't forget, in these very difficult puzzles, the order of operations applies.

Head on back to number 92 (if you haven't been there already) for a refresher if you need it!

71

The US SR-71 Blackbird spy plane ...

can travel about 71 kilometres in 71 seconds. That'll cut your commute time ...

Doubly-true (still!) alphametics

It's alphametic time again!

Now we're going to wrestle with FIVE + SIX + SIXTEEN + NINETEEN + TWENTY = SIXTYSIX:

```
        FIVE
         SIX
     SIXTEEN
    NINETEEN
+     TWENTY
------------
    SIXTYSIX
```

No direct mention of 71, but as I explained earlier, there aren't that many doubly-true alphametics going around so the tenuous link I rely on here is that we use 5 and 66 and 5 + 66 = 71.

If you're really flying with these, feel free to have a crack all by yourself. If you've still got your 'alpha-training' wheels on (and there's no shame in that) ... hit up the sidebar.

These are really hard, so I'll happily supply the hint that SIXTYSIX translates as 89140891.

This actually renders this alphametic fairly easy to solve, so off you go.

70

Turn it up! No, turn it *down*!

If you happen to be a 70-kilogram, 40-year-old man, you're in luck. Especially if full-blown 1960s-style business suits are your preferred attire.

You see, according to my old mate Dr Karl, for the past half-century or so, air conditioning in commercial buildings has been set *precisely* with you in mind.

The American Society of Heating, Refrigeration and Air Conditioning Engineers took the average office worker and calculated the optimal temperatures based on a range of specific factors — temperature and humidity, obviously, as well as metabolic rate and clothing worn by office workers to come up with a standard, Standard 55 ... back in 1966.

And their archetype? A 40-year-old man who weighed 70 kilograms and had to dress in a full business suit.

You probably don't need Dr Karl to tell you that not everyone in your average office nowadays is a 70-kilogram, 40-year-old male in a three-piece suit. Or that women have different metabolic rates and wear different clothing.

So what's considered the optimal range for your air con today? Glad you asked.

The Australian *Occupational Health and Safety Act, 2004* is in place to ensure that you're not cryogenically frozen at your desk, or indeed burned to a cinder. But the *Compliance Code for Workplace amenities and environment* helpfully informs us that 'optimum comfort for sedentary work is between 20°C and 26°C, depending on the time of year and clothing worn'.

I'm not going to lie, that's a fairly large range for you to argue about in your own office. But at least it allows some wriggle room for those of us non-40-year-old, 70-kilogram men in suits.

What about humidity? Too low, and it can cause respiratory problems. Too high and, well, ick.

Standard 55 (thankfully updated as recently as last year) also covers this.

Optimum humidity levels are considered to fall between 40–60%.

At 60% humidity in summer, the standard suggests a temperature range of 23–25.5°C.

And in winter, the same level of humidity would dictate a temperature range of 20–24°C.

Let the thermostat tussles begin!

Blankety blanks (#70)

Head on back to number 100 if you need a reminder of how these puzzles work. Rules at the right, pens at the ready ... take it away!

$$(\bigcirc + \bigcirc) \times (\bigcirc - \bigcirc) \times (8 \div (\bigcirc - \bigcirc) - 2) = 70$$

$$(\bigcirc + \bigcirc) \times ((\bigcirc + \bigcirc) \times (\bigcirc + \bigcirc) \div 8 - 2) = 70$$

$$(6 \times ((\bigcirc - \bigcirc) \div \bigcirc + \bigcirc) - \bigcirc) \times (\bigcirc + \bigcirc) = 70$$

$$(\bigcirc + \bigcirc) \times (\bigcirc + \bigcirc) \div (4 \times (\bigcirc - \bigcirc) - \bigcirc) = 70$$

$$((\bigcirc - \bigcirc) \times (\bigcirc - \bigcirc) - 6) \times (\bigcirc \times \bigcirc + \bigcirc) = 70$$

Reach the target number by filling in the blanks in each equation.

A completed equation must incorporate each and every digit from 0 to 9.

You cannot move the operations (+, −, ×, ÷) found between blanks.

Order of operations applies, and don't forget you have the digits 1, 2, 3, 4, 5, 6, 8 and 9 at your disposal.

70

The prôtostype

Ever wondered why we call prime numbers 'prime'? Well, over two millennia have passed since Euclid defined such numbers in his classic *The Elements* as:

'prôtos arithmos estin ho monadi monêi metroumenos'.

Come again? Roughly translated from the ancient Greek: 'prime numbers are that unit alone measured'. Or as we might nowadays say, *divisible*, rather than measured.

But what about this word, 'prôtos'? In Greek philosophy prôtos is used to express the first in the order of existence. So, 'prime', as we might say.

The word 'prime' itself comes from the Latin word *primus*, meaning 'first'.

In this sense, 'primes' are the first numbers, or the numbers from which all others are derived by multiplication. All positive integers can be measured by primes, but primes can be measured only by units.

The word prôtos continued to be used for generations, although it was by no means universal. Iamblichus, who died in 325 AD, liked the term 'euthymetric'. Thymaridas (350 BC) called them 'rectilinear', since they can only be represented one-dimensionally.

But when Sir Henry Billingsley first translated Euclid's *Elements* into English in 1570, he established 'prime' as the word of choice. It stuck.

Not bad, Sir Henry. Probably beats calling them TRULY AWESOME NUMBERS, as I might have done.

70

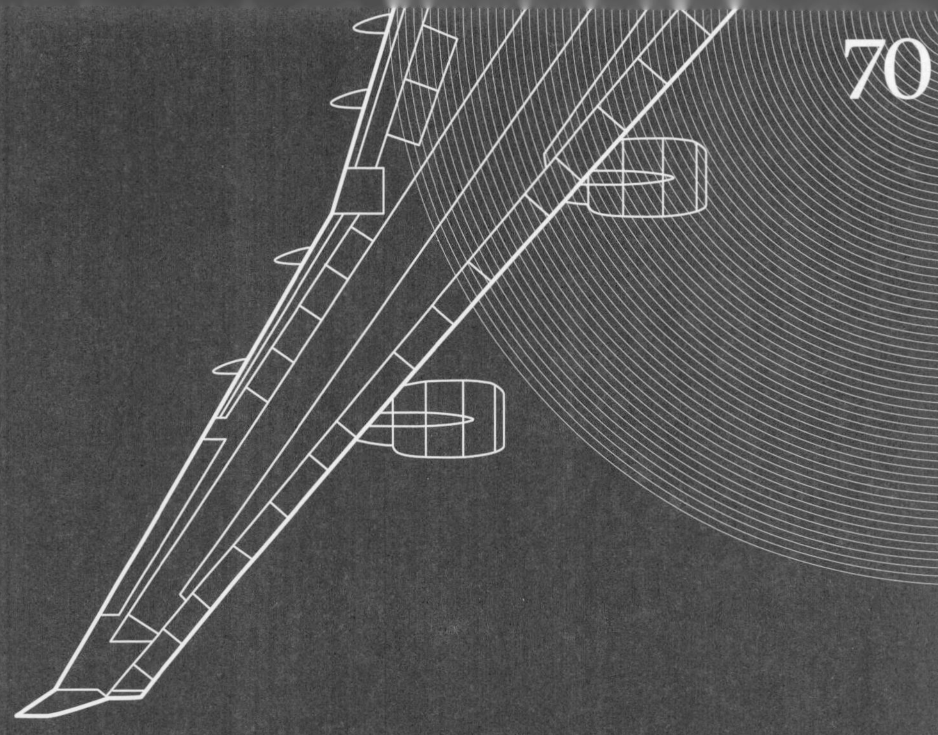

Boeing's 747 Jumbo Jet made its debut on 21 January 1970

Its first commercial flight, on this day, was flying Pan American's route between New York and London.

Carrying anywhere between 416 and 660 passengers (depending on the layout ... you guessed it, the 660 configuration is all cattle-class, all the time ...) the 747 held the passenger capacity record for an astonishing 37 years until it was beaten by the Airbus A380 in 2007.

69

Let's get dark ... real dark

We tend to think of the universe as being full of stuff. Just like the room I'm sitting in now is full of stuff — the table, that banana, the glass of water, even the air I breathe is still ... 'stuff'. There's just stuff everywhere ... isn't there?

But when we measure all the stuff we can see in the universe — not at the level of tables and bananas, but stars and planets and vast intergalactic clouds of dust — and when we *compare* that to other things we can measure in the universe, such as its size and the speed at which it is expanding, well that just doesn't make sense given how much stuff we can see.

So physicists believe there are two quantities that have an amazing impact on the universe, but about which we still know very little.

We call them 'dark energy' and 'dark matter'. The word 'dark' here means 'right now we can't detect them, but they seem to be there'.

Currently, the best estimates are that of all the 'stuff' in the universe, 69% is dark energy, helping the universe expand at an increasing rate.

A further 26% is dark matter, giving the universe mass even though we can't see it.

And that means that just 5% of the universe is matter that we can see — stars, interstellar clouds ...

And, of course, bananas.

My brilliant friend Ruben Meerman, the surfing scientist and one of Australia's great science educators, likes to say D.A.R.K. stands for 'Don't Actually Really Know'.

Sean's Syndesis

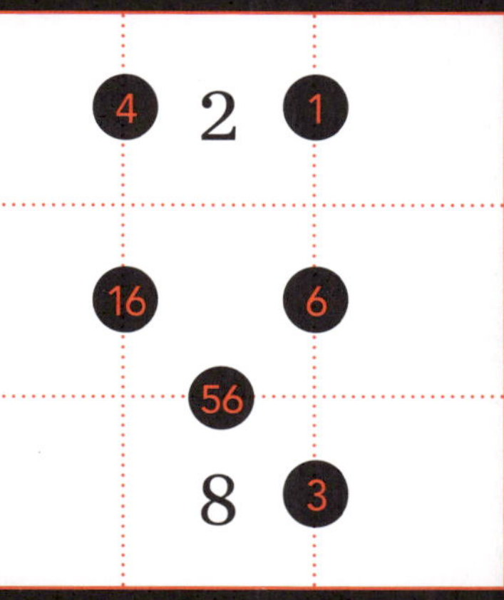

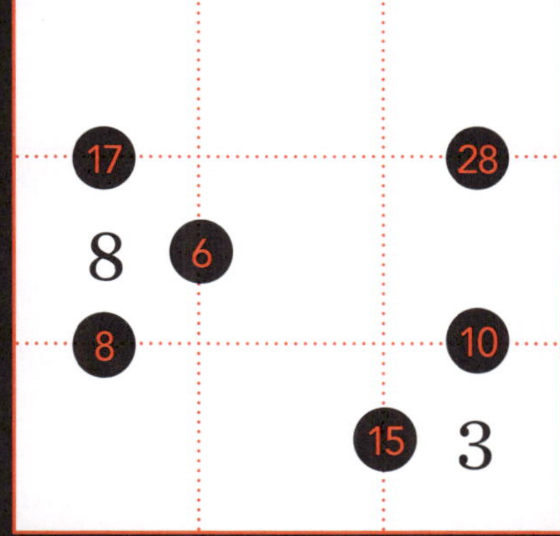

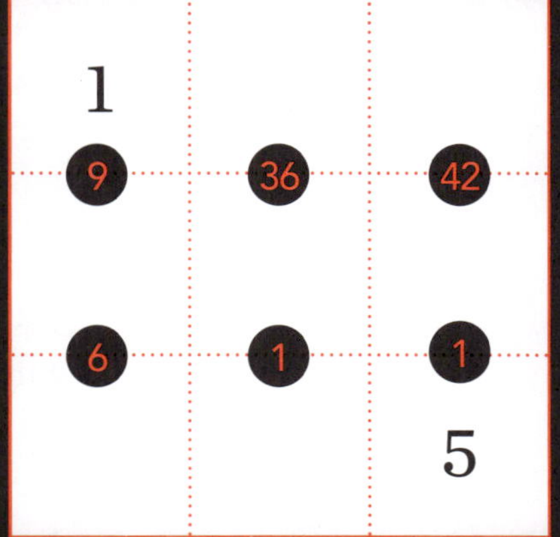

Enter the numbers 1 to 9 into each 3 × 3 grid so that each circled number is the result of adding, subtracting, multiplying, or dividing the two numbers in the cells it touches.

Head on back over to number 99 to read more about these awesome puzzles fresh from the mind of Sean Gardiner!

68

Space flies

It probably won't surprise you that if you fly off into the sky there isn't exactly a billboard sign saying 'Welcome to Outer Space'.

When it comes to defining where outer space begins, many scientists point to an imaginary line circling 68 miles (110 kilometres) above the Earth called the Karman Line.

While many people have heard of Laika the Russian 'first dog in space' (November 1957, aboard Sputnik 2), 10 years earlier just after World War II (20 February 1947 to be precise), the United States sent the very first animals into space.

Captured German V-2 rockets transported fruit flies into the skies (wow what a surprise) to study the effect of exposure to radiation at high altitudes.

It took just over 3 minutes for the V-2s to reach the Karman Line, making the fruit flies, whose names have sadly not been recorded for posterity, the first animals in space!

What we generally term 'space' is remarkably close overhead, at 110 kms.

For reference, Tasmania is 240 kms from mainland Australia.

Dancin' a jig

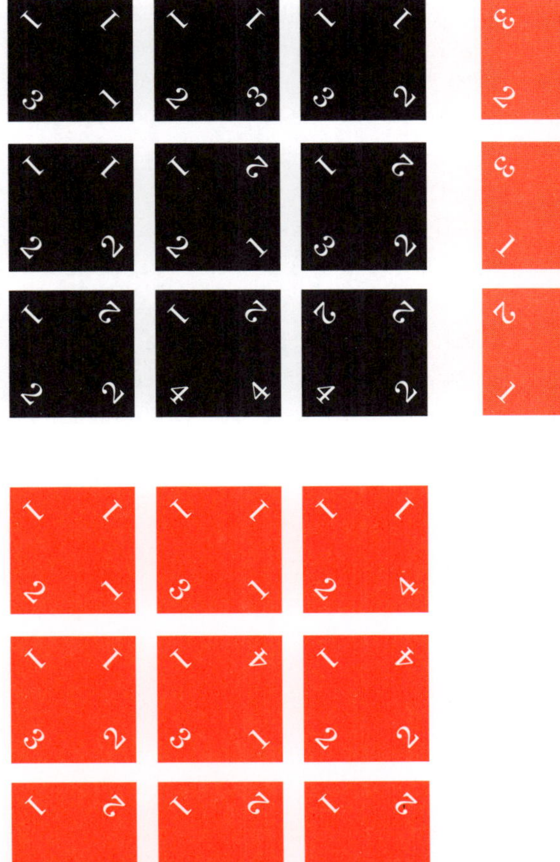

Arrange each set of 9 tiles into a 3 × 3 square so that adjacent tiles show the same numbers along their touching edges. Tiles may need to be rotated, but never reflected. Each set has only one solution. Check number 98 for more information.

For a downloadable cutout, head over to adamspencer.com.au/resources.

67

The Big D

Just in case you're wondering, 67 is the largest number to be raised to a 5th power in the equation $24^5 + 28^5 + 67^5 = 3^5 + 54^5 + 62^5 = 1,375,298,099$.

This sort of an equation, which features whole numbers raised to whole powers, is called a Diophantine equation after the ancient Greek mega-maths-man Diophantus of Alexandria.

I should point out you don't need all the powers to be the same. The observation that the only integer solution to $x^2 + 1 = y^3$ is $x = 0, y = 1$ is also a result concerning a Diophantine equation (as was the amazing mic-drop of an equation back at 73). In fact, Diophantine equations don't even need to include powers — they're just polynomial relations seeking only integral solutions. The most common example is $ax + by = 1$ for given a, b.

While the Big D examined these sort of equations in the 3rd century, it actually took until the 1900s before mathematicians really got on top of these beautiful but difficult creatures.

This isn't the only example of a number, in this case 1,375,298,099 being written as the sum of three 5th powers in two different ways. In fact there are an infinite number of them!

Here's a thorny little problem you might like to nut out. The next smallest example involves the numbers (13, 18, 44, 51, 64, 66). Can you work out which way these 6 numbers split into two groups of 3 such that when each is raised to its 5th power the two sets of 3 numbers have the same sum? And what is the sum?

I'll let you use a scientific calculator if you've got one handy, but in fact a smart phone should do the job for you. Another incredible fact about the world we live in IMHO.

Also, while you're here, why not try to match up the numbers on the side in their groups of 3, and find the total sum in these 3 other cases.

(8, 21, 43, 62, 68, 74)

(53, 56, 67, 72, 81, 83)

and ...

(39, 49, 75, 92, 100, 107)

67

New high score!

Welcome back to the game of High Scoring Equation where you fill in the blanks to win ~~real cash prizes~~ fame and glory! Have fun!

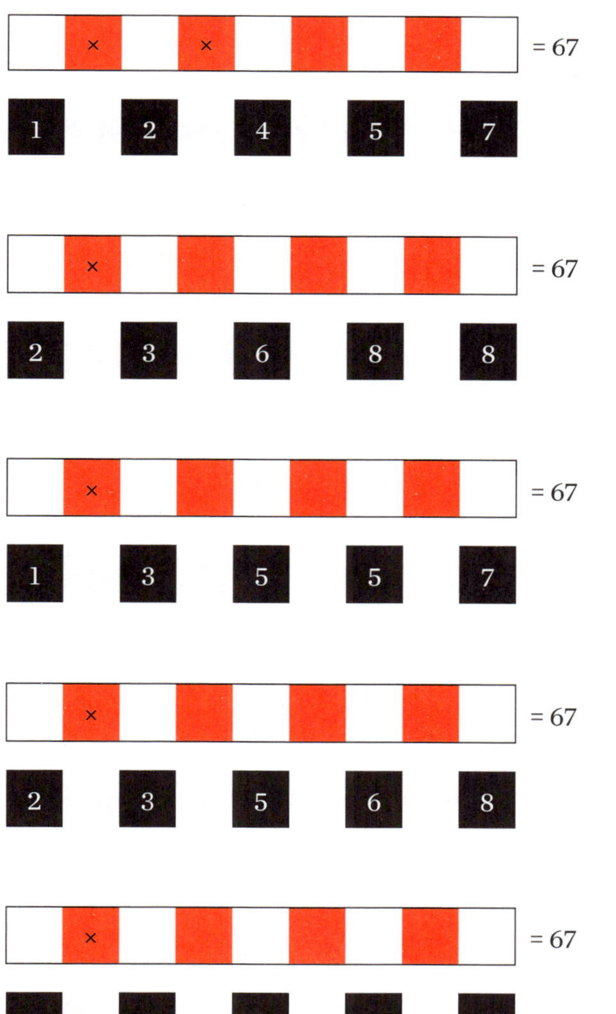

Reach the goal number by creating an equation using the provided numbers and your own choice of operators.

- The numbers should be placed in the white squares.
- Operations (+ – × ÷) are placed in orange squares.
- Order of operations matters, and you can't use brackets!
- Read the numbers off in order for your final score.
- Your goal is to find the equation that gives the highest score.

Head back over to number 97 if you need a refresher on these, otherwise ... get cracking!

67

Comet comet comet comet comet chameleon

It's easy to take things for granted in this amazing age. The mobile phone you hold in your hand puts you one minute away from more information than the smartest people in the world had access to just a generation ago.

I'm using the same sort of phone to write this paragraph from a café in New York (namedrop!) while waiting for a friend.

I just used that same phone to FaceTime my daughter who this year received an injection of Gardasil, a vaccination against the HPV virus that will hopefully help us conquer the disease cervical cancer.

So yeah, it's easy to get blasé when you hear of yet another stunning achievement in this era of amazement.

But occasionally I like to pause and remind myself that in my lifetime, humanity landed a robotic device called Philae ... on a comet.

That's right ... *on* a comet.

It all started on 2 March 2004 when the European Space Agency launched the Rosetta space probe. Named after the the ancient Egyptian stone that was covered in 3 languages and led to us translating ancient texts, the probe and its lander Philae spent 10 years travelling to the comet

67

67P also known as Churyumov–Gerasimenko after the two Russian scientists who discovered it.

Rosetta swung out past Mars, then off to the Kuiper belt around Jupiter where it studied the comet, just over 4 kilometres long and wide, as it whizzed along at speeds of around 135,000 km/h.

Then, on 6 August 2014, it moved into orbit around the comet and on 12 November dispatched the lander Philae which *landed on the freakin' comet*!

After briefly sending back data from the surface of 67P, Philae stopped talking to Rosetta and then, poetically, Rosetta crash dived onto the surface of 67P on 30 November 2016.

A truly remarkable engineering feat.

Honestly, I've got to say I prefer the name Churyumov–Gerasimenko.

Then again, if it wasn't also called 67P it wouldn't feature here in the countdown!

66

America's famous Route 66 was originally numbered Route 60

The highway famously ran from Chicago through Missouri, Kansas, Oklahoma, Texas, New Mexico and Arizona before ending in California – a huge 3940 kilometres.

Although originally keen to secure the round number '60' as the designation of their spanking new highway in the 1920s, officials had to settle on the number '66' after delegates from Kentucky kicked up a fuss, wanting a highway from Virginia to LA to be known as 'Route 60'.

The rest as they say is history. 'Get your kicks on Route 66' certainly has a better ring to it than 'Route 60'.

Pretty hexy!

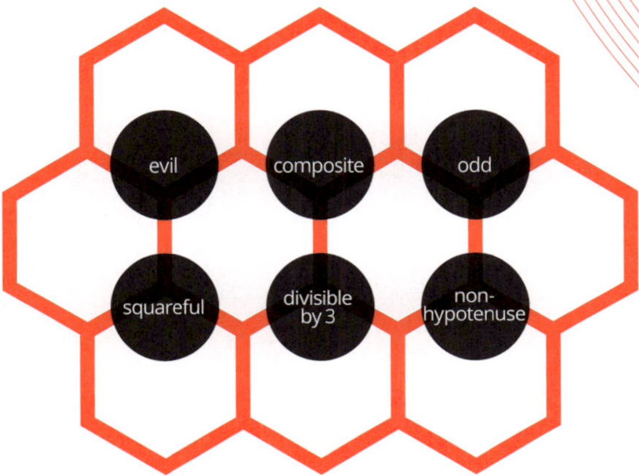

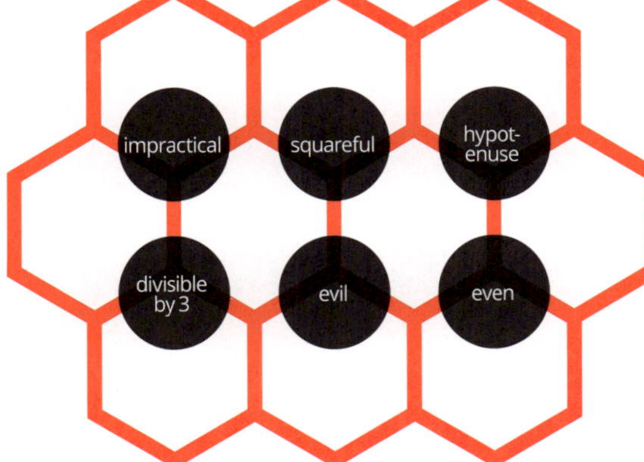

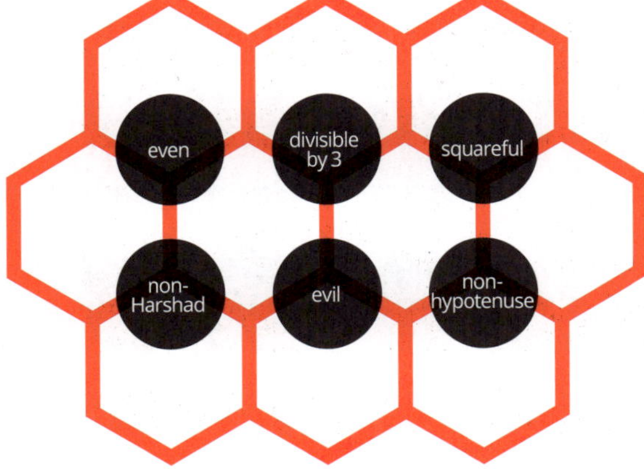

Place the numbers 61 to 70 into each hexagonal grid to satisfy all the rules.

Each cell is touching one or more categories in circles. The number in each cell must belong to all the categories touching it.

Need a hint?

Don't forget the definitions to all of these categories are at very end of the book and that the 6-hex is probably the easiest ... so feel free to start there, and work up to 66.

Winner, winner, chicken (65) dinner

Chicken 65 is a spicy, deep fried snack originating in Chennai, India. It was first introduced by A.M. Buhari of the Buhari Hotel chain in 1965, which may give us a clue about its name.

Further evidence of Mr Buhari's preferred naming conventions might be found in his later dishes, Chicken 78, Chicken 82 and Chicken 90 which were introduced in 1978, 1982 and – go on, guess – that's right: 1990 respectively.

Surely we're just about due for a Chicken 2018?

65*

The good people @qikipedia love putting oddities out there on Twitter.

I highly recommend a trawl through their page. But be warned ... it's a rabbit warren of astonishing facts and you'll get stuck reading. Plan accordingly.

Those mischievious maths-lovers certainly caught my eye when they pointed out this little gem.

Say you had 100 guests attending a wedding.

You seat them all 10 guests per table (that's '10 pax' to all you hospitality folk ...). Simple.

How many ways do you think the room could be seated?

I hope you're sitting down for this one. If you don't care about the order of the tables within (so if they are not themselves numbered) and you don't care who sits where at each table, there are 65 trillion trillion trillion trillion trillion trillion *trillion* seating possibilities!

Fair to say no matter how great your guest list, that's a lot of zeroes ...

* trillion trillion trillion trillion trillion trillion trillion ...

65

Haul glass

Now, I like to avoid a queue at the counter as much as the next guy (#efficiency).

So ordering a few drinks or bites to eat for your friends as well as yourself makes sense across a period of hours.

But when it comes to carrying a few orders at once, none of us is in the same league as German stein-packer Oliver Struempfel, who on Monday 4 September 2017 in Abensberg Bavaria successfully carried 29 glasses of beer the required 40 metres to break his own world record of 27 steins!

You're not just born with this ability. 'The Struempf' had been training 4 times a week in the gym for months before hauling glass.

And I should point out, as impressive as 65 kilos of beer in 29 steins is, he actually started with 31 steins and smashed two along the way.

I hope he wasn't trying to serve beer to 30 mates because someone is going to miss out!

A one-litre stein of beer — the usual object of the 'Masskrugstemmen' Bavarian beer-carrying contest — weighs approximately 2.25 kilograms.

I'll leave you to crunch the numbers for The Struempf's hefty feat as you tote your next pint away from the bar ...

On the tiles

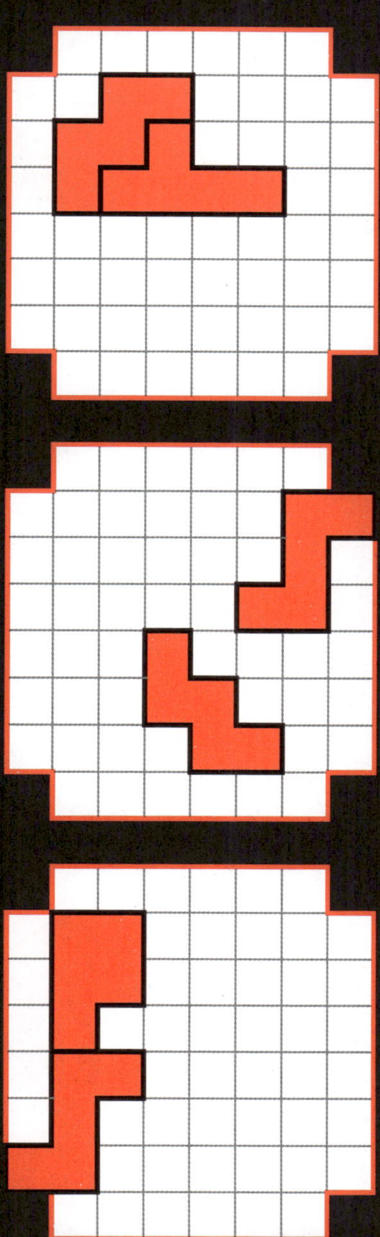

In each of these grids, finish tiling it so that it contains one of each pentomino type.

The 12 different pentominoes are shown here.

You may need to rotate and/or even flip some of them to get them to fit.

Head to adamspencer.com.au/resources if you'd like to download a copy of the puzzle to cut up!

64

$$16 \times 4 = 64$$
$$166 \times 4 = 664$$
$$1666 \times 4 = 6664$$
$$16666 \times 4 = 66664$$
$$166666 \times 4 = 666664$$
$$1666666 \times 4 = 6666664$$
$$16666666 \times 4 = 66666664$$
$$166666666 \times 4 = 666666664$$

Maths + coffee

Like a cup of coffee first thing in the morning?

Boffins from the US Army have developed an algorithm that allows them to determine, from person to person, 'when and how much [each] individual should take caffeine to achieve peak performance at the desired time, for the desired duration'.

Their study suggested that the algorithm helped improve alertness by 16–64%, while reducing caffeine consumption by 17–65%. Win!

Unlike some of the earlier, ah, more notorious US military research projects featuring real people, they calculated their findings using existing computer models to predict schedules, sleep deprivation and caffeine impact alertness during different times of the day.

So say you're gearing up for that big exam — this algorithm would help you determine the best time to slam your strong double piccolo latte for maximum alertness.

You might need a strong coffee before you attack the algorithm, though. The 'Optimisation problem to find a caffeine-dosing strategy that minimises neurobehavioural performance impairment' goes something like this:

$$\min Z = 50\frac{AUC_C(t_i, D_i)}{AUC_{NC}} + 50\ \frac{WP_C(t_i, D_i)}{WP_{NC}} + 250 \max\{C\ (t_i, D_i) - C_{\max}, 0\}$$

The conditions are on the right. (You'll have to read the report yourself for the constraints, I'm afraid: onlinelibrary.wiley.com/doi/full/10.1111/jsr.12711)

If you don't fancy crunching the numbers, the algorithm is being incorporated into the site 2B-Alert, a free online tool that 'predicts alertness of an "average" individual'.

Stay tuned and check it out at 2b-alert-web.bhsai.org.

'where Z denotes the objective function that we wish to minimise. The optimisation variables t_i and D_i represent the time (in hours after the first wake-up time in the schedule) and the caffeine amount (in mg) of dose *i*, respectively, with $i = 1, 2, \ldots, n$ (the number of doses). *AUC* denotes the area under the PVT mean response time (RT) curve above the baseline, and *WP* the difference between the peak of the mean RT curve and the baseline. $C(t_i, D_i)$ denotes the level of caffeine in the blood. The subscripts *C* and *NC* denote caffeine and no caffeine, respectively. The last term penalises the objective function when the maximum value of $C(t_i, D_i)$ is higher than the maximum caffeine level achieved by a single dose of 400 mg, denoted by Cmax.'

Phew! Make mine a double.

64

The Beatles' hit 'When I'm Sixty-Four'...

was recorded in 1966 in the key of C major.

Paul McCartney, who sings the lead vocal, suggested to (legendary producer) George Martin that he speed up the master take to raise the key by a semitone and make his voice appear younger.

The result, aside from (in McCartney's words) making the song 'more rooty-tooty', was that the final release was in the key of D-flat major.

Personally, I would have called it 'When I'm 8^2'.

I think that's why Sir Paul is the billionaire!

64

Beat the gridlock!

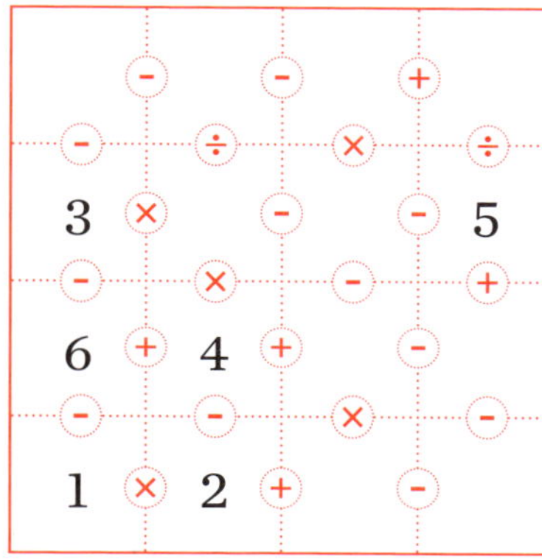

=6

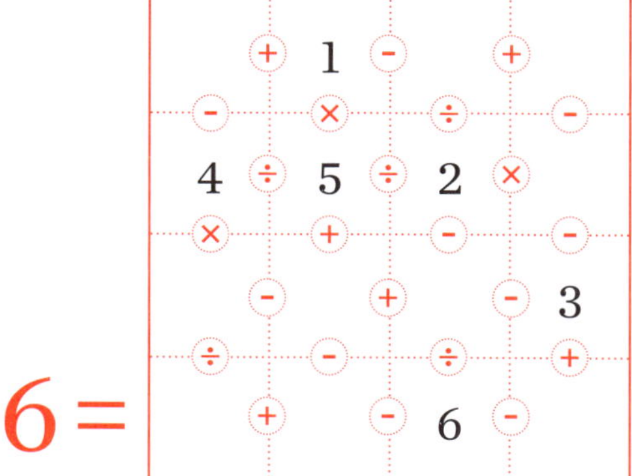

	÷	6	÷	5	×	
–		×		–		+
	–	2	×	4	÷	
–		–		×		–
	–		–	3	+	
+		+		+		×
	+		–		×	1

=6

The rules for these are easy. The answers, well, they're a little tougher!

For each 4 × 4 grid, enter the numbers 1 to 16 so that each row and column equals the target number. As always, order of operations matters.

Can you break the gridlock?

If you don't like defacing this book, go to adamspencer.com/resources for a template to fill in.

63

Ekl!

Having just explained what a Diophantine equation is, why not meet another wonderful example?

Turns out, thanks to the work of R.L. Ekl, (I know – cool name!) that:

$$63^9 + 54^9 + 51^9 + 49^9 + 38^9 + 35^9 + 29^9 + 24^9 + 21^9 + 12^9 + 10^9 + 7^9 + 2^9 + 1^9 = 66^9.$$

This is the smallest example there is of a 9th power being the sum of fourteen 9th powers.

If you'd rather have a 9th power as the sum of only twelve 9th powers, our good old buddy Jean-Charles Meyrignac (what is it with mathematicians with cool names?) showed in 1997 that:

$$91^9 + 91^9 + 89^9 + 71^9 + 68^9 + 65^9 + 43^9 + 42^9 + 19^9 + 16^9 + 13^9 + 5^9 = 103^9.$$

This leads to an obvious question, 'can we express a 9th power as exactly thirteen 9th powers'?

Well, my curious friend, as of today ... we simply don't know.

I'll keep you posted.

If you hear anything, please get straight back to me.

Drill bits (day 4) ...

Today the players are being led by #63 Australian Womens' cricket sensation Ashleigh Gardner who, in June 2017, took the field against Pakistan and became the first indigenous Australian woman to play in the world cup of cricket.

In which order did the players stand around the circle?

- #63 stood at the top of the circle.
- #3 and #93 stood the same distance from #33.
- There were 2 people between #73 and #53.
- #83, #3, and #13 stood together in some order.
- From #23's perspective, #3 was opposite them.
- There were 2 people between #43 and #33.
- #3 stood closer to #43 than to #33.
- #63, #83, and #53 stood together in some order.
- From #73's perspective, #13 was on the left side of the circle.

Ten players stood in a circle for their daily drill.

They are numbered 3, 13, 23, 33, 43, 53, 63, 73, 83 and 93.

Each day a different player is the drill leader, which by delightful coincidence, corresponds to the chapter number and 'front position'.

Your task, coach, is to work out the order they stood in the circle each day given the list of clues provided by these fun-loving sporty puzzlers.

Game on!

62

‘In 2015, just 62 individuals had the same wealth as 3.6 billion people — the bottom half of humanity’

This figure is down from 388 individuals as recently as 2010, according to Oxfam, who included this data in their annual report.

The figure was met with some scepticism given the huge disparity — so much so that ABC Fact Check looked into it.

Depressingly, they found that Oxfam’s claim is in the ballpark.

62

Telesto, Calypso, Tethys, oh my!

Most scientists agree that Titan is Saturn's most interesting moon.

This is because methane, which on Earth we find mostly as a gas in natural gas and smelly dad farts, is present on Titan as a liquid, a solid and a gas. Much like water on Earth we think it exists in lakes, evaporates into the sky and rains back down.

Some scientists think this 'methane cycle' makes Titan a prime candidate to be the first place in the universe outside of Earth to support some form of life. That's pretty awesome.

So, well, you need something cool to be Saturn's most gnarly moon. After all, Saturn has 62 of them. Saturn has so many moons we've only gotten around to naming 53 of them.

Three other really cool moons are Telesto and Calypso which orbit Saturn but do so stuck in a gravitational lock with another moon called Tethys.

As the moon, say Tethys (at L_2 in the diagram on the right), orbits the central planet, in this case Saturn, two Trojan moons are locked in at points L_4 (Telesto) and L_5 (Calypso) and therefore also orbit Saturn.

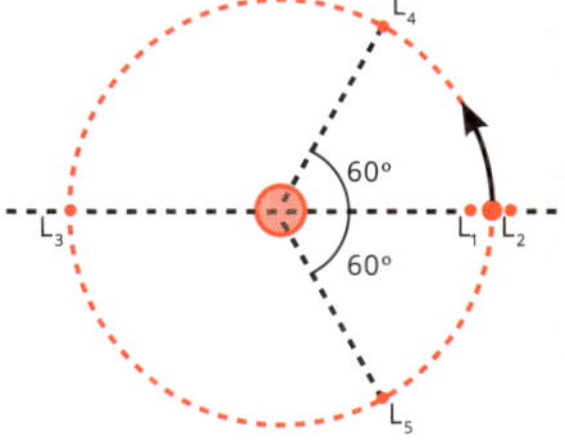

These points L_4 and L_5 are called Lagrange points after the 18th century French maths and physics super gun Joseph-Louis Lagrange. Lagrange points are where a smaller body is effectively 'trapped' by the gravitational effect of two or more larger bodies.

Saturn also provides us with the only other known Trojan moons in the solar system, with Helene and Polydeuces locked in sync with Dione as it orbits the big S.

62

Oymya-goodness! That's cold

The coldest regularly-occupied place on Earth, where people live full-time, is probably Oymyakon, a remote Siberian outpost in Russia's Yakutia region.

The town was once a stopover for reindeer herders in the 1920s and 30s who took water from its hot thermal spring that did not freeze. The ground in Oymyakon is so cold that there are no pipes (toilets are outdoor holes in the ground) and to break the earth — say to build a new toilet or bury someone — you have to first light a bonfire to thaw the frozen surface.

The 500 residents of Oymyakon, where school stays open if it's only in the minus 40s, regularly leave their cars running even if they are not using them to prevent them freezing up and not restarting. It's *that* cold!

'Cold?' you say ... exactly *how* cold?

In January 2018, during the middle of winter (when it is dark for up to 21 hours a day), a particularly cold snap hit and the fancy new digital thermometer that had been installed to showcase how famously cold it can get broke as it registered –62°C.

In 2013, presumably with a tougher old-school thermometer, the village recorded a cool –72°C.

Snakes alive!

→

1	+	5	+	1
–	6	–	7	–
9	+	1	×	7
–	6	×	6	=
4	+	7	=	62

→

7	+	9	–	1
×	6	+	4	×
5	×	8	–	1
+	8	–	8	=
7	–	9	=	62

→

7	+	8	–	6
–	6	×	8	×
2	+	1	–	8
+	2	–	5	=
1	+	3	=	62

In these grids, you can make paths from the top-left corner to the bottom-right, by moving between adjacent squares.

You can move up, down, left or right, but you can never visit the same square twice.

And, you guessed it, the paths must trace out the correct equation to reach the target number in the bottom right-hand corner.

For each grid, there are 3 different paths that all trace out correct equations.

Head on back to number 92 (if you haven't been there already) for a refresher if you need it!

61

F*k*!

If you've read any of my previous books (all available at adamspencer.com.au and good bookstores #shamefulplug) then you're probably familiar with the famous Fibonacci sequence.

Well, here is something (else) pretty cool about this famous list of numbers.

Write out the first, say, 20 terms of the sequence. Remember the sequence is formed by starting with 1, 1 and then adding the most recent two numbers to get the next number. So 1 + 1 = 2, 1 + 2 = 3, 2 + 3 = 5 and so on:

> 1, 1, 2, 3, 5, 8, 13, 21, 34, 55, 89, 144, 233, 377, 610, 987, 1597, 2584, 4181, 6765.

You might need a glass of water after that!

But check this out. The 4th term of the sequence is $F_4 = 3$. Let's look at the multiples of 4, that is 8, 12, 16 and 20. $F_8 = 21$, $F_{12} = 144$, $F_{16} = 987$ and $F_{20} = 6765$ and these numbers are all multiples of $F_4 = 3$.

Similarly, $F_5 = 5$ and $F_{10} = 55$, $F_{15} = 610$ and $F_{20} = 6765$ are all multiples of 5.

Can you see the pattern here?

For any whole number *n*, it seems that the Fibonacci number $F_{(n \times k)}$ is a multiple of F_k.

In fact, this is the case.

So by looking at our list you can see that every 15th Fibonacci number will be a multiple of 61.

Doubly-true (still!) alphametics

Time for another little DTA (which I'm sure absolutely no one calls these things — so let's get this trending now!).

And for the purists, we actually involve the exact number of this page in the equation this time.

I'd like you to solve:

```
        SIX
       NINE
     ELEVEN
    SIXTEEN
+  NINETEEN
-----------
   SIXTYONE
```

Once again, the letters all correspond to unique digits. The equation holds, in word form, when the digits are subbed in. None of the leading letters can be equal to 0, and if you can work it out without any help you are truly crushing this!

If you'd like a hint — and there's never any harm in asking — in this puzzle, SIX translates as 926.

61

61 is the country code when dialling Australia from around the world

That should be reasonably well known to you. But if you'd like a little bit of perhaps-lesser-known numerical phone trivia, try this.

Within Australia, if you're on your mobile and about to call someone who you rather *didn't* have your number, you can fumble around in:

> Settings > General > Um > Back out of General, go into > Phone > Damn it! *Where is it?!* >>>>>

... or, to save all that trouble, just dial #31# before the number of the person you wish to keep in the dark.

From one of those quaint, archaic, fixed-line phones dial simply 1831.

#TooEasy

I ♥ cuban primes

Mathematicians love prime numbers and have spent so much time studying them they have found all manner of types.

You've got palindromic primes that read the same way forwards and backwards like 101; you've got Fermat primes which are numbers of the form:

$$F_n = 2^{2^n} + 1$$

... where F_n is prime. For example:

$$F_3 = 2^{2^3} + 1 = 257.$$

Well, 61 is a *cuban* prime. This does not mean it likes laying on a beach in Varadero and dancing the Pachanga. It is called cuban because its definition involves cubes.

A prime number p is a cuban prime if it satisfies either:

$$p = \frac{x^3 - y^3}{x - y}, x = y + 1, y > 0 \text{ or } \ldots$$

$$p = \frac{x^3 - y^3}{x - y}, x = y + 2, y > 0$$

The cuban primes less than 100 are 7, 13, 19, 37 and 61.

Question time. Can you find the values of x and y that generate these cuban primes?

60

One league = one hour's walk

Although it's no longer an 'official' unit of measurement in any nation, the word originally referred to the distance a person could walk in an hour.

Let's be honest, you'd have to think that people like Usain Bolt travel a wee bit faster than the likes of, say, me. It's fair to say such elite striders are ... wait for it ... in a league of their own. (Sorry.)

A dodecahedron

An icosahedron

Why you good-fer-somethin', two-bit, 60-diagonalled Platonic solid ...

A regular dodecahedron or pentagonal dodecahedron is a dodecahedron that is, well, regular. It's composed of 12 regular pentagonal faces — 3 meeting at each of its 20 vertices and 30 edges.

You might remember from such wonderful, bestselling tomes as *Adam Spencer's Big Book of Numbers* (PLUG!) that it's also one of the 5 Platonic solids. So is an icosahedron, which has 20 faces, 12 vertices and 30 edges.

So far, so good. But here's the hard part.

Which would fit best in a sphere — a dodecahedron or an icosahedron? That is, if you fit the largest possible dodecahedron and icosahedron inside a given sphere, which leaves the least space?

Honestly, the maths here is very difficult.

Instead, just try to picture each shape in a sphere and hazard a guess as to which fills more space.

Even this is not easy to do!

Both shapes have the same number of edges. One has more faces, the other has more vertices. Does that help you make a decision?

Good luck.

Blankety blanks (#60)

Head on back to number 100 if you need a reminder of how these puzzles work. Rules on the left, pens at the ready ... go for it!

$$(\bigcirc + \bigcirc - \bigcirc)\times((\bigcirc - \bigcirc)\div(5 - 3) + \bigcirc) = 60$$

$$(\bigcirc + \bigcirc)\times((\bigcirc - \bigcirc)\times(\bigcirc - \bigcirc - \bigcirc) + 1) = 60$$

$$(\bigcirc \times((\bigcirc + \bigcirc)\div(\bigcirc + \bigcirc) + \bigcirc) - \bigcirc) \times 5 = 60$$

$$(((\bigcirc - \bigcirc) \times \bigcirc + \bigcirc) \times \bigcirc + 3)\div(\bigcirc + \bigcirc) = 60$$

$$(\bigcirc + \bigcirc)\times(\bigcirc - \bigcirc)\times(8 \div (\bigcirc - \bigcirc) - \bigcirc) = 60$$

Reach the target number by filling in the blanks in each equation.

A completed equation must incorporate each and every digit from 0 to 9.

The answer already contains the 0 and 6 so you are allocating the digits 1, 2, 3, 4, 5, 7, 8 and 9 to the circles.

You cannot move the operations (+, −, ×, ÷) found between blanks.

Order of operations applies!

Infinite pieces of pi

One of the most amazing of all numbers, possibly the most important and beautiful of all, is pi (π).

To obtain π, compare how far it is around a circle (the circumference) with how far it is across (the diameter).

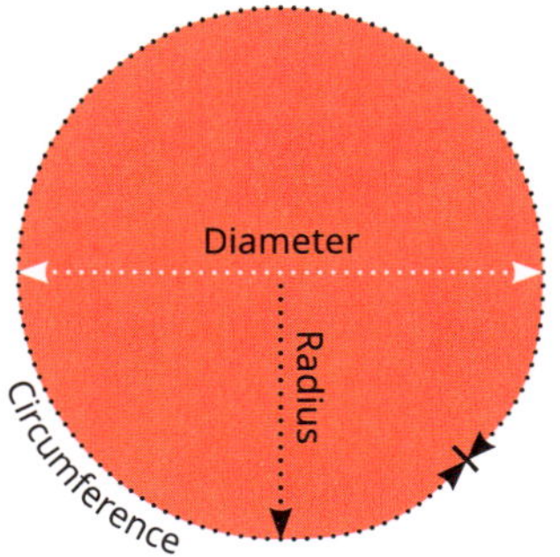

If you do this with a piece of string around, say a dinner plate, you'll see that π is 'a little bit more than 3'. As a rough decimal approximation you could use $\pi \approx 3.14$.

But I'm sure most of you know that π is *not* equal to 3.14. It is not equal to 3.14159 or even 3.14159265, even though these are both more accurate decimal expansions of this magical quantity.

In fact, the the decimal expansion of π goes on forever. It doesn't stop and never gets to a point after which it just repeats. It is, as we like to say, an infinite non-recurring decimal number.

60

We have calculated π to trillions of decimal places (I know – mathematicians get really seriously into some things, don't they!), far enough to observe all sorts of beautiful bits of trivia.

The first double of a number comes with 33 appearing at decimal places 24 and 25, followed soon after by an 88 and a 99. The first triple is the 111 that starts at decimal place 153.

If you're looking for the string of digits 123456789, you'll be looking for a while, it starts at position 523,551,552; and if it's 987654321 you're after you'll have to keep chugging all the way to the 719,473,323rd decimal place.

If you're looking for the 10-digit string 0123456789 you go well beyond the mere hundreds of millions of decimal places you've already combed through and don't find happiness until the 17,387,594,880th decimal place.

But if you don't care about the order of the 10 digits you first find them in a row as early as decimal place 60.

Look, here they are: 3.14159265358979323846264338327950288419716939937510582097494**4592307816**40628620899862803.

You might have heard of the tiny, cheap computer called a Raspberry Pi.

Its name is reportedly derived from the creators wanting to continue the seemingly inexplicable tradition the tech world has with fruit (Tangerine Computer Systems, Blackberry and Apple to name just a few ...)

And the 'Pi' bit? Well, it's not an homage to the Great Number itself, but an acronym, standing for 'Python [a programming language] Interpreter'.

Keep that one up your sleeve for your next round of pub trivia with the fellow geeks!

We have only found about 60 'stellar mass' black holes in our entire galaxy

At least so far. With nothing able to escape its clutches, not even light, it's not easy to spot a black hole ... that's the whole point (get it!).

But if a black hole has a companion star dancing around it that helps greatly. These stellar mass black holes are formed by the gravitational collapse of a massive star, say from 5 times up to hundreds of times the size of our sun.

Sixty black holes might sound like a lot, but given how hard most are to see and how little of the galaxy we've examined, it is believed that the supermassive black hole in the centre of the Milky Way is surrounded by, get this, around 10,000 other black holes!

59

Fun with 59

The number 59 is a rocking little number that features in some lovely mathematics.

For a start it's a prime number.

But not just any prime number. You can construct a 3 × 3 magic square entirely of primes with 59 in the middle. Recall from number 78 that the rows, columns and diagonals of a magic square all have the same sum.

In this case, I'll give you a hint.

The rows, columns and diagonals all sum to 177. Can you find the other 8 primes to complete this magic square?

	59	

59

Sean's Syndesis

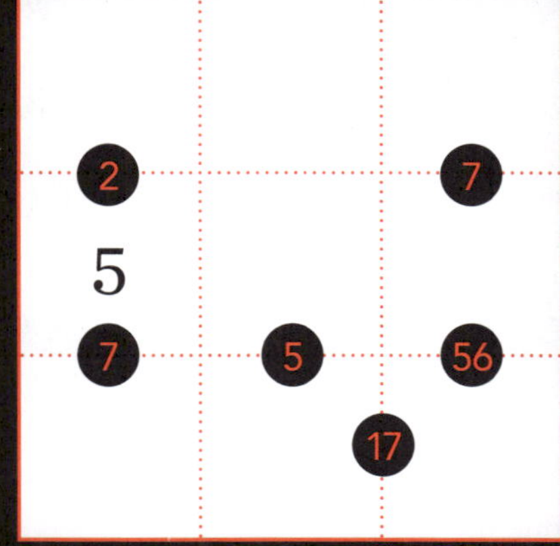

Enter the numbers 1 to 9 into each 3 × 3 grid so that each circled number is the result of adding, subtracting, multiplying, or dividing the two numbers in the cells it touches.

Head on back over to number 99 to read more about these awesome puzzles fresh from the mind of Sean Gardiner!

Stellated by starlight

This rather striking little shape is one of the 59 stellations of the icosahedron. Say what? 'Stellation' means the extension of the faces of a polyhedron until they intersect with the extensions of other faces, but in such a way that the symmetry of the original polyhedron is preserved.

The enumeration of all 59 stellations of an icosahedron was not accomplished until 1938 when H.S.M. Coxeter, P. Du Val, H.T. Flather and J.F. Petrie published their page-turning, barnstormer of a book *The Fifty-Nine Icosahedra*. Nice one.

Further 59 fun ...

Further 59 fun involves a class of numbers called 'primorials'. We have already met the factorials back at number 100. Remember that 'five factorial' is written as $5! = 5 \times 4 \times 3 \times 2 \times 1$. This is equal to the number of ways we can arrange the 5 letters ABCDE (for example ABCED, BCDAE and so on).

The primorial of a prime number, which uses the # symbol, is the product of that number with all the primes less than it. So $5\# = 5 \times 3 \times 2$ for example.

Numbers of the form $p\# + 1$ are called Euclid numbers after the great Greek mathematician Euclid. We call them such because he used numbers of this form in his beautiful proof that there are an infinite number of primes.

Euclid reasoned thus: let's say there are a finite number of primes, 2, 3, 5 ..., p where p is the largest prime number. Consider the number $p\# + 1$. Clearly $p\# + 1$ is not divisible by p because $p\#$ is and we'd have a remainder of 1 when we divided $p\# + 1$ by p. But similarly dividing $p\# + 1$ by 2 or 3 or 5 or any other prime up to p leaves this remainder of 1.

So $p\# + 1$ must either be a prime number itself, or divisible by a prime that was not on our original list. Either way our list of 'all the primes' was not complete. There can never be such a finite list of primes. The primes must therefore be infinite!

Played, Euclid.

So where does 59 fit into all of this? It turns out that the first few Euclid numbers are themselves prime. $2\# + 1 = 3$; $3\# + 1 = 3 \times 2 + 1 = 7$; $5\# + 1 = 5 \times 3 \times 2 + 1 = 31$ and so on.

Now it's time to strap your thinking cap on. The first composite Euclid number is divisible by 59. Find this Euclid number and factorise it.

58

Za raditeley!

It was a mere 58 cm wide polished metal dish ...

But it ushered in one of the most exciting decades in all human history.

Sputnik 1, from the Russian *Prosteyshiy Sputnik* or 'elementary satellite' (I know, Prosteyshiy Sputnik sounds so much better!) carried 4 external radio antennas which broadcast a pulse signal that could be detected, even by amateurs, back on Earth.

On 4 October 1957 the Soviet Union launched the 83 kg Sputnik into a low Earth orbit and, over the next 3 weeks, it orbited over almost the entire inhabited planet giving off its pulse.

Once the batteries died it lasted another couple of months before falling back to Earth.

Within a year, in response to the publicity and perceived scientific superiority of the Soviets, the US had formed NASA and we were off to the (space) races.

It's easy to think that because they got to the moon that the US 'won' the space race, but the Russkis were doing incredible stuff at the same time.

Check out the documentary *Cosmodrome* for starters.

58

Spinnin' a jig

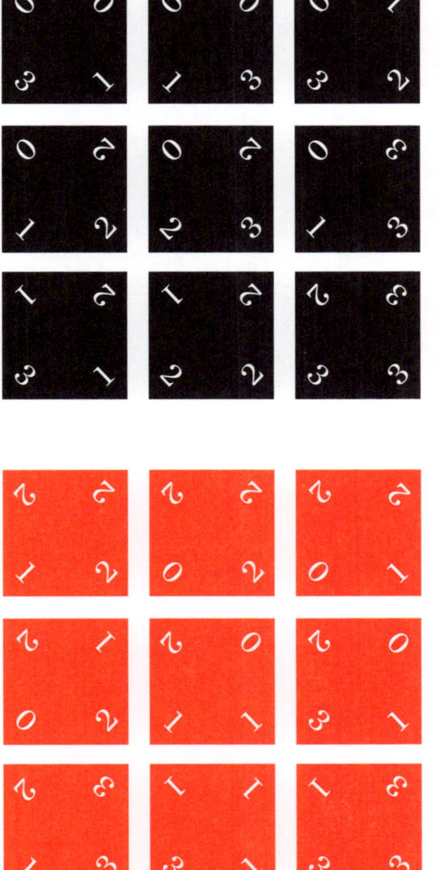

Arrange each set of 9 tiles into a 3 × 3 square so that adjacent tiles show the same numbers along their touching edges. Tiles may need to be rotated, but never reflected. Each set has only one solution. Check number 98 for more information.

For a downloadable cutout, head over to adamspencer.com.au/resources/

Naughty Newton

Sir Isaac Newton was one of the most influential thinkers in human history.

He made incredible discoveries throughout his life, but to put his brilliance into perspective, in a period of just two years while the Great Plague tore through England, Newton pioneered a new theory of light and optics, discovered and quantified gravitation and developed a revolutionary new approach to mathematics which we call infinitesimal calculus.

Sheer, unadulterated genius.

But in many ways he was also a product of his times. Newton was a deeply religious man who set incredibly high standards for himself and others when it came to what he thought was acceptable behaviour that would keep his God happy.

He set such high standards he once famously wrote down a list of all the sins he had committed either side of Whitsunday – the 7th Sunday after Easter and a significant date on the Christian and Anglican calendars.

The year in question was 1662, Isaac was 19 and, by his own admission, he got up to at the very least, the following sinful activities, reproduced here in all their olde English glory:

Sins* committed before Whitsunday 1662

* By that I mean activities he, Newton, felt were sinful. I am making no judgement here.

1. Using the word (God) openly
2. Eating an apple at Thy house
3. Making a feather while on Thy day
4. Denying that I made it
5. Making a mousetrap on Thy day
6. Contriving of the chimes on Thy day
7. Squirting water on Thy day
8. Making pies on Sunday night

9. Swimming in a kimnel on Thy day
10. Putting a pin in Iohn Keys hat on Thy day to pick him
11. Carelessly hearing and committing many sermons
12. Refusing to go to the close at my mothers command
13. Threatning my father and mother Smith to burne them and the house over them
14. Wishing death and hoping it to some
15. Striking many
16. Having uncleane thoughts words and actions and dreamese
17. Stealing cherry cobs from Eduard Storer
18. Denying that I did so
19. Denying a crossbow to my mother and grandmother though I knew of it
20. Setting my heart on money learning pleasure more than Thee
21. A relapse
22. A relapse
23. A breaking again of my covenant renued in the Lords Supper
24. Punching my sister
25. Robbing my mothers box of plums and sugar
26. Calling Dorothy Rose a jade
27. Glutiny in my sickness
28. Peevishness with my mother
29. With my sister
30. Falling out with the servants
31. Divers commissions of alle my duties
32. Idle discourse on Thy day and at other times
33. Not turning nearer to Thee for my affections
34. Not living according to my belief
35. Not loving Thee for Thy self
36. Not loving Thee for Thy goodness to us

57

37. Not desiring Thy ordinances
38. Not long [longing] for Thee in [illegible]
39. Fearing man above Thee
40. Using unlawful means to bring us out of distresses
41. Caring for worldly things more than God
42. Not craving a blessing from God on our honest endeavors
43. Missing chapel
44. Beating Arthur Storer
45. Peevishness at Master Clarks for a piece of bread and butter
46. Striving to cheat with a brass halfe crowne
47. Twisting a cord on Sunday morning
48. Reading the history of the Christian champions on Sunday

Since Whitsunday 1662

49. Glutony
50. Glutony
51. Vsing Wilfords towel to spare my own
52. Negligence at the chapel.
53. Sermons at Saint Marys (4)
54. Lying about a louse
55. Denyhing my chamberfellow of the knowledge of him that took him for a sot
56. Neglecting to pray 3
57. Helping Pettit to make his water watch at 12 of the clock on Saturday night

For all I know, Wilford may have been completely cool about Isaac using his towel? Maybe it was only a white lie about the louse? Certainly a lot of his self admonishment rotated around keeping the sabbath holy, while I, for one, am absolutely fine with you twisting a chord or helping a mate build a water watch on a Sunday.

Now, there's no way to sugar-coat this, Sir Isaac, calling Dorothy Rose a jade was a bit rough.

But for what it's worth, I reckon you probably did more good than bad, on balance.

New high score!

Welcome back to the game of High Scoring Equation where you fill in the blanks to win ~~real cash prizes~~ fame and glory! Have fun!

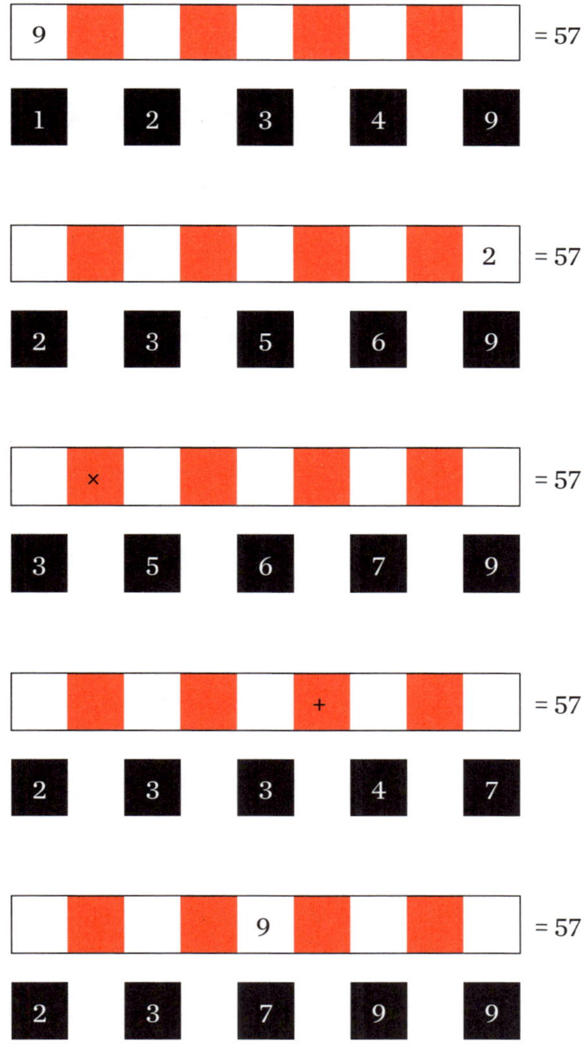

Reach the goal number by creating an equation using the provided numbers and your own choice of operators.

- The numbers should be placed in the white squares.
- Operations (+ – × ÷) are placed in orange squares.
- Order of operations matters, and you can't use brackets!
- Read the numbers off in order for your final score.
- Your goal is to find the equation that gives the highest score.

Head back over to number 97 if you need a refresher on these, otherwise ... get cracking!

Latin squares

Hey, did you know that there are 56 normalised 5 × 5 Latin squares? *Quid dicis*?

Latin squares are square grids of size $n \times n$ containing n elements in each row and column, with no element repeating in a row or column. So on the left we have a 3 × 3 Latin square with the 3 elements A, B, C; but the grid on the right is not a 4 × 4 Latin square because the second and fourth columns have repeat emojis in them.

A	B	C
C	A	B
B	C	A

A Latin square is called 'normalised' if the first row and column have the elements in the same order. So notice that any Latin square can be normalised. In the 3 × 3 case

A	B	C
C	A	B
B	C	A

and

B	C	A
A	B	C
C	A	B

reduce to

A	B	C
B	C	A
C	A	B

If the elements are letters or numbers, we would normally write the normalised square with elements in the first row and column in alphabetical or numerical order. Harder to do in the case of emojis! The grid on the right is the *only* normalised 3 × 3 Latin square.

Turns out there are 56 normalised 5 × 5 Latin squares. I'd be mighty impressed if you found all of them. Why not try to find all 4 normalised 4 × 4 Latin squares instead?

More hexes!

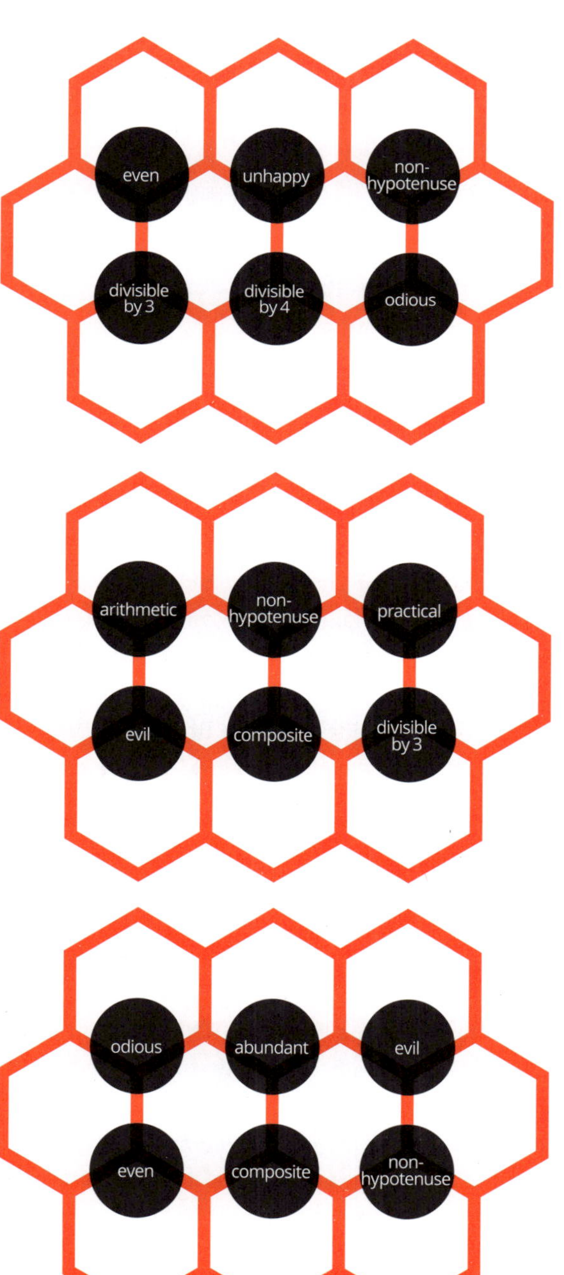

Place the numbers 51 to 60 into each hexagonal grid to satisfy all the rules.

Each cell is touching one or more categories in circles. The number in each cell must belong to all the categories touching it.

Need a hint? If the definitions are a bit tough for you, there's a full list of what these words mean at the end of the book.

55

Respect overdue

In March 2018 *The New York Times* made a confession. It was to do with the obituary section of the paper. The obits record the lives of people who have died. Politicians, sports stars, movie idols and regular hard-working individuals who have made the world a little bit better.

While no one was suggesting that any individual who'd had an obit published hasn't deserved it, when you looked back across the 167 years of *NYT* obituaries, 'the vast majority chronicled the lives of men, mostly white ones'.

So the *Times* launched 'Overlooked', an obituary series recognising 'remarkable women who did not receive obituaries when they died (going all the way back to the 1850s). Some were pioneers in their fields yet virtually unknown to the wider world; others were famous to some degree either before or after their deaths but still didn't make the cut.'

The article singled out how Anemona Hartocollis, a *Times* reporter, faced the challenge of composing a tribute to the brilliant but tortured poet Sylvia Plath that would be running 55 years after her death.

And none to soon!

Graphic content

One of the subjects covered in higher mathematics is called graph theory. How would you like to learn a little bit about it?

Okay, let's go!

A graph is a set of points joined by lines which we call edges, like the image on the right. This shows a graph with 6 points and 7 edges.

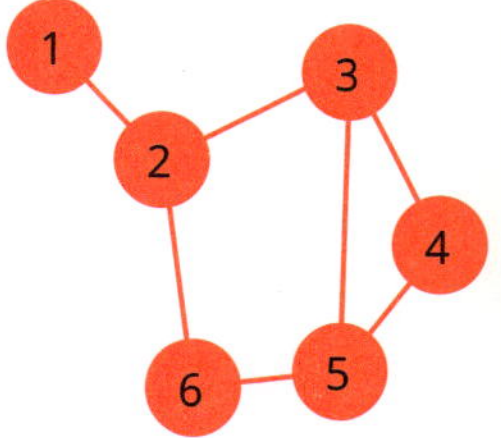

A graph is *simple* if it has no loops nor multiple edges between the same two vertices. So, out of the three graphs below, the graph on the left is simple, but the other two are not.

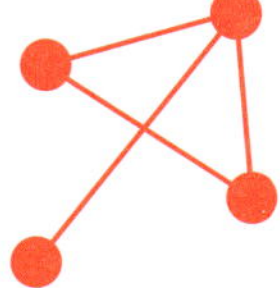

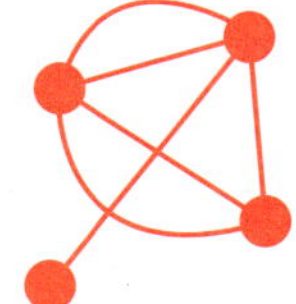

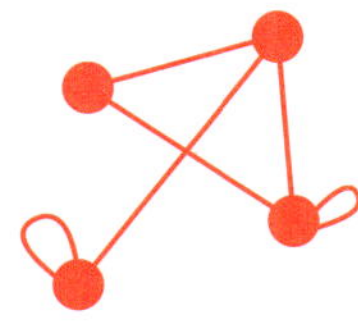

Graphs are *undirected* if a line from point A to point B means the same thing as a line from B to A. So if the points were people and a line meant 'A shakes hands with B' then the graph of handshakes is undirected.

If the line from A to B meant 'A is older than B', then the graph of age relationships is called *directed,* because it matters which way the line runs.

55

Finally, (for us here anyway -- trust me when I tell you that graph theory is *massive*!) a graph is complete if every pair of points is joined.

So, this graph, K_5:

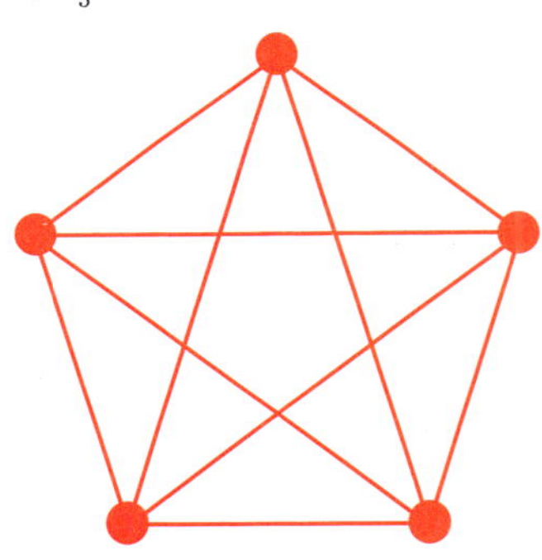

... is the first complete graph you've seen here.

The complete graphs with n points, or 'of order n' are labelled K_n like K_5 here.

Okay, with that now under your belt, try this little exercise on for size. A few minutes ago, it would have been gobbledy-gook, but now I think you can land it.

Draw the complete, simple, undirected graphs K_2, K_3, K_4 and K_5.

Count the number of edges for each of these graphs. Can you see a pattern? For which graph does K_n have 55 edges?

Leonhard Euler's paper on the Seven Bridges of Königsberg, published in 1736, is regarded as the first paper in the history of graph theory.

You can find out more about that online or ... by checking out the brilliant, amazing, must-have collector's item *Adam Spencer's Big Book of Numbers* by some guy I know ...

On the tiles

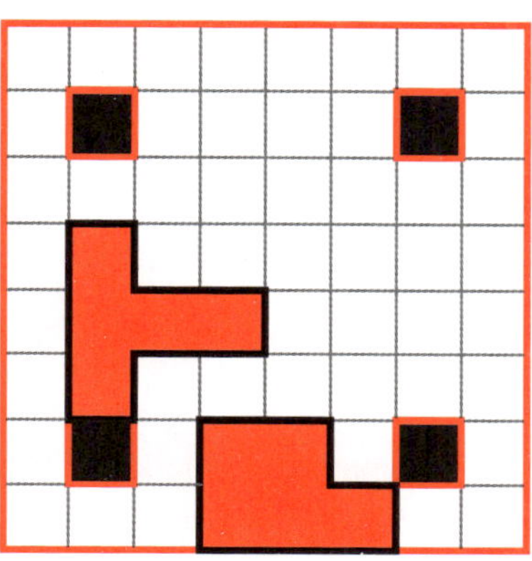

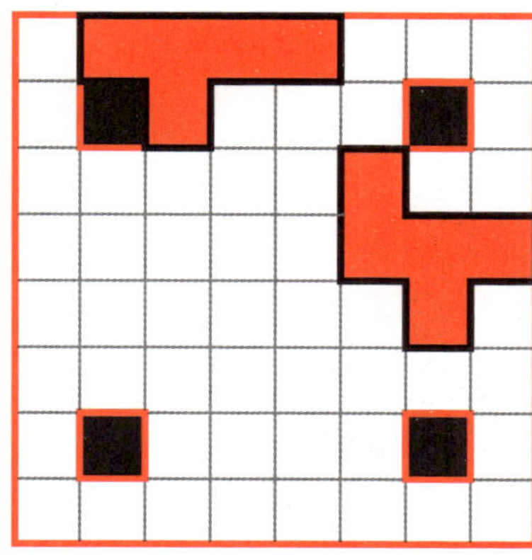

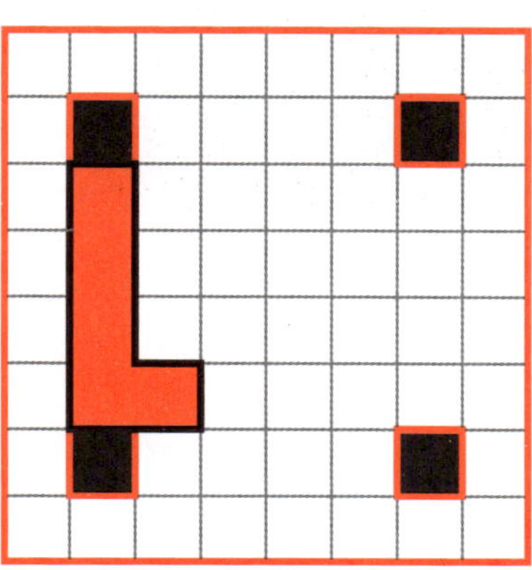

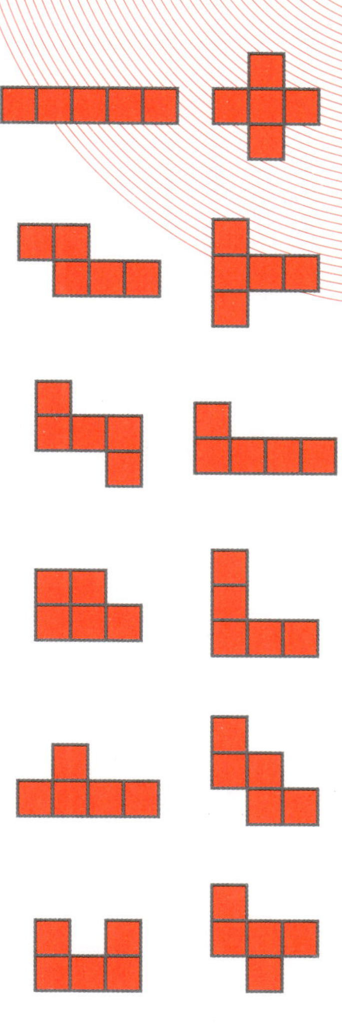

In each of these grids, finish tiling it so that it contains one of each pentomino type.

The 12 different pentominoes are shown here.

You may need to rotate and/or even flip some of them to get them to fit.

Head to adamspencer.com.au/resources if you'd like to download a copy of the puzzle to cut up!

Tasty ...

Here are 3 fantastic geometry puzzles for you that come from the book *Geometry Snacks* by Ed Southall and Vincent Pantaloni.

Good luck – these start alright but get pretty tough pretty quick!

Let's kick things off with this equilateral triangle. It has area 54 cm². The line across the middle hits the two sides exactly halfway up. What is the area of the triangle that is shaded red?

The great thing about the puzzles in *Geometry Snacks* is that they can be solved in more than one way.

Try and solve these, and if you really enjoy doing them, maybe you can find a second (or third!) way to get to the answer.

The next one's a bit harder.

Four semicircles with radius 54 cm are constructed in the square on the opposite page.

Tell me, what is the area of the square?

Did you find a way to solve both of these so far? Maybe more than one way if you're really crushing it!

Finally, in diagram 3 we have an outer boundary that is a quarter of a circle of radius 54 cm. There are two semicircles inside and a circle. All these internal shapes are tangent to each other. What is the radius of the circle and smaller semicircle?

Cubetastic

If you're the type of person who likes expressing a cube as the sum of 3 cubes ... and, hey, who isn't? ... then you'll just love the number 54.

Not only does $54^3 = 12^3 + 19^3 + 53^3$ and $22^3 + 51^3 + 54^3 = 67^3$ but $7^3 + 54^3 + 57^3$ is *also* a cube.

Can you work out this last one?

Extra points if you can do it by hand!

54

The gridlocker

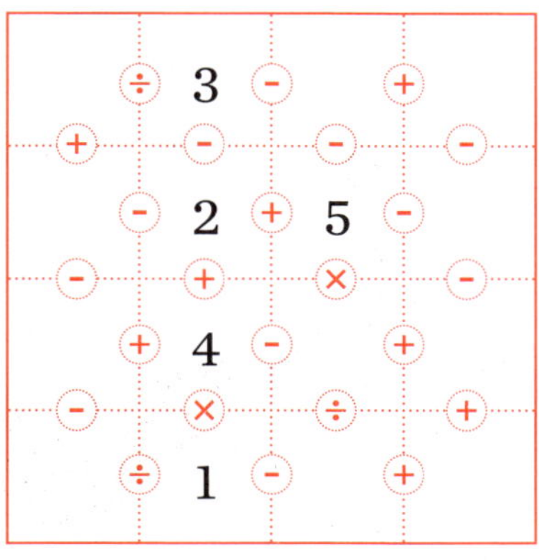

= 5

5 =

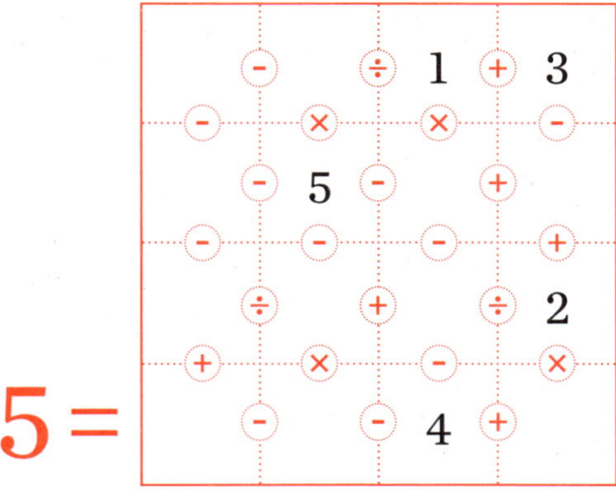

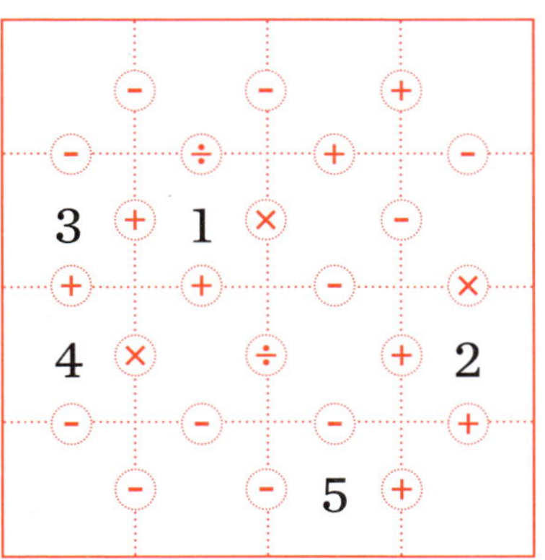

= 5

The rules for these are easy. The answers, well, they're a little tougher!

For each 4 × 4 grid, enter the numbers 1 to 16 so that each row and column equals the target number. As always, order of operations matters.

Can you break the gridlock?

If you don't fancy defacing this book, go to adamspencer.com/resources for a template to fill in.

53

Julie Amiri, a British housewife, was detained for shoplifting 53 times between 1985 and 1993

She is often referred to online, not entirely unfairly, I'd say, as the greatest nuisance to English police of all time.

In 1993 in Chichester Crown Court she argued that the reason she committed these acts, at an average of once every couple of months for almost a decade, was that the thrill of being chased by the police aroused her sexually. It was, in fact, the only way she could achieve such happiness.

She was not convicted.

That's right — she, ahem, got off in court.

Drill bits (day 5) ...

Today's drill will be taken by #53 Major League Baseball Hall of Fame LA Dodger's pitcher (and three time World Series Champion) Don Drysdale ...

In which order did the players stand around the circle?

- #53 stood at the front of the circle.
- #53, #83, and #43 stood together in some order.
- From #73's perspective, #93 was opposite them.
- #63 stood closer to #73 than to #33.
- #93, #13, and #23 stood together in some order.
- From #23's perspective, #13 was on the right side of the circle.
- #83 stood next to #63.
- #3 stood closer to #23 than to #63.
- #53, #63, and #83 stood together in some order.
- From #3's perspective, #53 was on the right side of the circle.

Ten players stood in a circle for their daily drill.

They are numbered 3, 13, 23, 33, 43, 53, 63, 73, 83 and 93.

Each day a different player is the drill leader, which by delightful coincidence, corresponds to the chapter number and 'front position'.

Your task, coach, is to work out the order they stood in the circle each day given the list of clues provided by these fun-loving sporty puzzlers.

Game on!

52

It hertz

Cosmos Magazine ran an excellent article in January 2018 entitled 'Five sounds science can't explain'.

One of them was a whale call, measured by navy equipment near the North Pole.

What makes this call perplexing is that while blue whales make a call at between 10 and 39 Hz (hertz) and fin whales at a frequency of 20 Hz, this whale song resonates at 52 Hz — much, much higher than those common calls.

Could it be a blue whale with something wrong with it? Or perhaps the baby of a blue and fin whale? We simply don't know.

Whatever the case may be, the 52 Hz call is unique in the oceans.

The lowest frequency us humans can hear is around 20 Hz, however we can 'feel' sounds much lower — an earthquake, for instance.

Snakes alive!

→				
6	×	9	×	9
+	5	–	6	–
2	+	4	×	2
–	2	×	8	=
1	–	2	=	52

→				
9	+	9	–	2
+	2	×	6	+
8	×	8	×	9
–	3	–	9	=
6	×	8	=	52

→				
6	×	6	–	9
+	7	×	9	+
8	×	8	–	9
–	5	–	8	=
1	–	6	=	52

In these grids, you can make paths from the top-left corner to the bottom-right, by moving between adjacent squares.

You can move up, down, left or right, but you can never visit the same square twice.

And, you guessed it, the paths must trace out the correct equation to reach the target number in the bottom right-hand corner.

For each grid, there are 3 different paths that all trace out correct equations.

Head on back to number 92 (if you haven't been there already) for a refresher if you need it!

52

Every day I'm shufflin'

As you know, I regularly sing the praises of John D. Cook who bombards the internet with nerdy maths brilliance from his numerous Twitter handles, blog posts and the like.

On 23 January 2018 he posed this question:

Take two decks of cards and shuffle them separately. Then place the decks face down next to each other. Turn the top card over on each deck at the same time. Are they the same card or different? Do the same for the second card in each deck – same or different? Keep going down through the deck. How many times out of the 52 cards in each deck would you expect the card in deck A and the corresponding card in deck B to be the same card?

There are two ways you might want to tackle this.

You could calculate the probabilities, or you could do the experiment by shuffling a couple of decks. But it should be clear to you that you can't just shuffle the two decks, say, 3 or 4 times.

You would have to be thorough and do the experiment a lot of times to guard against getting a freak result that really skews your answer.

And when it comes to being thorough, John D. Cook doesn't muck around.

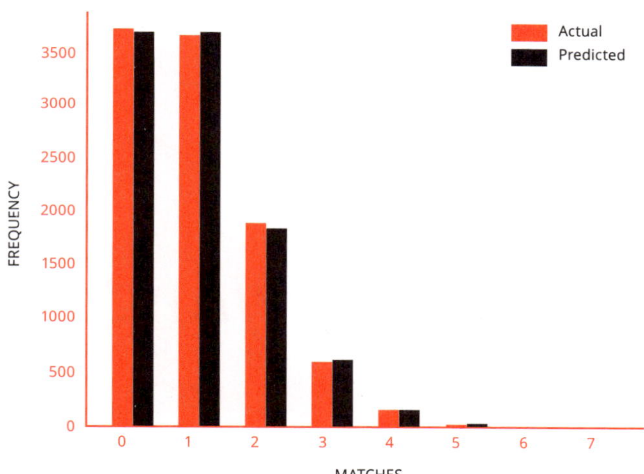

Rather than buying two packs of cards and tediously going through the process, with the chance of making a mistake from tiredness or, let's be honest, sheer boredom, John wrote an algorithm in the Python coding language that simulated 10,000 shuffles of two 52-card decks*.

The actual results came out remarkably close to what was predicted by probability calculations. About $1/3$ of the time you get zero matches and it's about the same for one match. The odds of 3 or more cards matching up is pretty small.

In fact, this $1/3$ figure is, to be exact, $1/e$ where e is the famous mathematical constant Euler's number and about equal to 2.71828. So you'd expect zero or one match in your decks to each occur about 36.7879% of the time.

* Want to play along at home? You can find John's Python code on his blog at johndcook.com/blog/2018/01/23/

51

It's a hap*pen*ing thing ...

If you're a serious pen lover ... cue some dads reading this to roll back like Homer Simpson contemplating a doughnut while their kids ponder 'what's a pen?' ... you would have heard of what is probably the most famous pen of all time, the Parker 51.

The 51 was so respected for its sleek design and performance (I know, they got very excited about pens back in the day) that Generals Dwight Eisenhower and Douglas MacArthur each used one when they signed the treaties that ended World War II.

Eisenhower even held the two 51s in a V for Victory sign which, as an accidental marketing ploy, was pure genius.

The 51 became synonymous with all things good, and though it set you back anywhere from $12.50 to $50 – up to $1000 today – sales of the Parker 51 broke the 2 million mark in 1947.

That's a lot of pen love.

Doubly true (still!) alphametics

More alphametics now.

If you've battled through the solutions with hints so far, you should be getting closer to being able to solve one all by yourself. I certainly hope so, because I don't want to alarm you, but this is your LAST HINT for an alphametic in this book!

Try this little doubly-true alphametic out, in honour of the number 51.

```
        THREE
         NINE
          TEN
     FOURTEEN
+     FIFTEEN
-------------
     FIFTYONE
```

Your hint, if you choose to accept it (and again, I'm totally cool if you do — these are really hard without the hints) is that TEN translates as 726.

Get alphameticing (I know, I know, not a word).

50

In 2018 the price of a standard single item of mail in the United States rose 1c to 50c

But not for former Presidents.

In addition to a pension, staff costs and 10 years of Secret Service protection, as a way of saying 'thanks mate', Obama, Clinton, Bush et al, all get free mail ... for life!

Blankety blanks (#50)

Head on back to number 100 if you need a reminder of how these puzzles work. Rules at the right, pens at the ready ... take it away!

$$(\bigcirc - \bigcirc) \times (\bigcirc - \bigcirc) \times (\bigcirc - (\bigcirc + \bigcirc) \div 2) = 50$$

$$(\bigcirc - \bigcirc) \times (\bigcirc - (\bigcirc + \bigcirc) \div (\bigcirc - \bigcirc)) \times 2 = 50$$

$$((\bigcirc + \bigcirc) \div 2 - \bigcirc) \times (\bigcirc + \bigcirc) \times (\bigcirc - \bigcirc) = 50$$

$$(\bigcirc + \bigcirc - \bigcirc) \times (7 - (\bigcirc - \bigcirc) \div (\bigcirc + \bigcirc)) = 50$$

$$((\bigcirc + \bigcirc) \times (\bigcirc + \bigcirc \div (\bigcirc - \bigcirc)) + \bigcirc) \div 2 = 50$$

You're down to 5 clues now, so after this you'll start getting versions of this puzzle in all their glorious difficulty!

Reach the target number by filling in the blanks in each equation.

A completed equation must incorporate each and every digit from 0 to 9, with the 0 and 5 already featuring here in the answers.

You cannot move the operations (+, −, ×, ÷) found between blanks.

Order of operations applies!

50

You say potato, I say ... *you're busted, punk!*

The internet isn't just about cat pictures and the Kardashians.

Some of the other enduring memes in this digital age are generated by people sifting through ancient legal statutes that no one ever bothered to take off the law books as things evolved and pointing out that these laws hilariously still apply.

It may surprise you to learn that a lot of these are furphies.*

Turns out, for instance, that there is no law against entering the state of Minnesota with a duck on your head! (I know. Trust me. Look online and the claim is out there.)

Nor are you banned from swimming at certain Australian beaches if you're not covered from neck to knee.

But as far as I can ascertain, due to the power residing in the Western Australian Potato Marketing Corporation, they can *still* enforce a law from 1946 that makes it illegal for an individual to be in possession of more than 50 kilos of spuds.

* The word is generally thought to come from the cast iron water carts made by John Furphy of J. Furphy & Sons of Shepparton, Victoria in the 1880s.

Many of Furphy's carts were used by the Australian Army during World War I and became popular as gathering places to exchange gossip and yarns — literally water cooler discussions — often of dubious veracity.

Thus the term 'furphy' was born.

50

Getting eban

Okay, I've got to level with you.

There are two pretty strong hints on this page for the following puzzle.

Here goes.

Which number follows in this sequence?

2, 4, 6, 30, 32, 34, 36, 40, 42, 44, 46?

Now — spoiler alert — I don't want to give too much away, but if you guessed 50 purely by looking at the chapter number, then it'd be hard to argue with you.

But before you go patting yourself on the back, tell me *why* that number is the correct answer?

One of the most famous teams in the NFL is the San Francisco 49ers

You might think the team's nickname derives from the year 1949 when they joined the NFL from the now defunct All-America Football Conference ... but you'd be wrong.

The 49ers are actually named after the prospectors who arrived in Northern California in the 1849 Gold Rush.

49

Banksy was here

Sometimes in mathematics, someone proves something, but then someone else improves upon that piece of knowledge.

Here's an example and it involves the number 49.

In 2015, American mathematician William D. Banks showed that every positive integer can be written as the sum of at most 49 palindromic numbers.

For example, we can write 389 = 141 + 88 + 44 + 44 + 44 + 11 + 9 + 8 as the sum of 8 palindromes (numbers which read the same when written in reverse).

Banks showed there is no number that needs 50 or more palindromes to be expressed in a sum like this.

But then, as sometimes happens, a year later, a pair of mathematicians, the Spaniard Javier Cilleruelo and his Romanian buddy Florian Luca, dropped a result that improved somewhat on old Banksy's observation.

C&L showed that any positive integer can be written as the sum of just 3 palindromes.

So we could have just written 389 = 11 + 55 + 323.

And the new result is even more beautiful than just slashing 49 down to 3, but to understand why, you have to understand the concept of the base of a number system.

We count in base 10. That means we have 10 digits, namely 0, 1, 2, 3, 4, 5, 6, 7, 8 and 9, but once we have counted up to 9 we run out of digits, so we get to 10. When we count all the way up to 19 we have again run out of digits to place after the 9, so we go to 20. Eventually at 99 we tick over to 100. After we hit 999 we tick over into the thousands and so on.

So when we write the number 2481, we can think of it as 2 thousands plus 4 hundreds plus 8 tens plus 1 or $2 \times 10 \times 10 \times 10 + 4 \times 10 \times 10 + 8 \times 10 + 1$.

Now, just say we only had 4 digits, 0, 1, 2 and 3.

Counting up from 0 we would go 0, 1, 2, 3, 10, 11, 12, 13, 20, 21, 22, 23, 30, 31, 32, 33, 100, 101 and so on. This is counting in base 4. In base 4 the number 22313 is the same as the number $2 \times 4 \times 4 \times 4 \times 4 + 2 \times 4 \times 4 \times 4 + 3 \times 4 \times 4 + 1 \times 4 + 3 = 2 \times 256 + 2 \times 64 + 3 \times 16 + 1 \times 4 + 3 = 695$ in good old base 10.

Or you might be familiar with counting in base 2, which we call binary. This counting system 0, 1, 10, 11, 100, 101, 110, 111, 1000, 1001 ... underpins computer coding and our modern digital world.

When Cilleruelo and Luca improved Banks's result from 49 palindromic numbers to just 3, they further showed that this happens not just in our familiar base 10 system. A whole number can be written as the sum of 3 or less palindromes in any base 5 or greater.

To sum things up in a fittingly palindromic word: WOW!

Sean's Syndesis

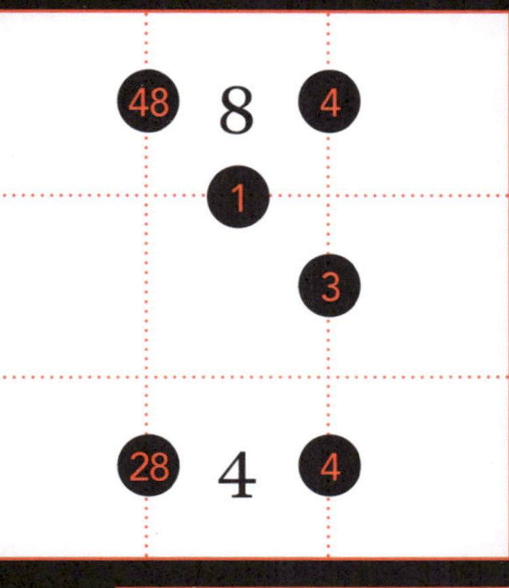

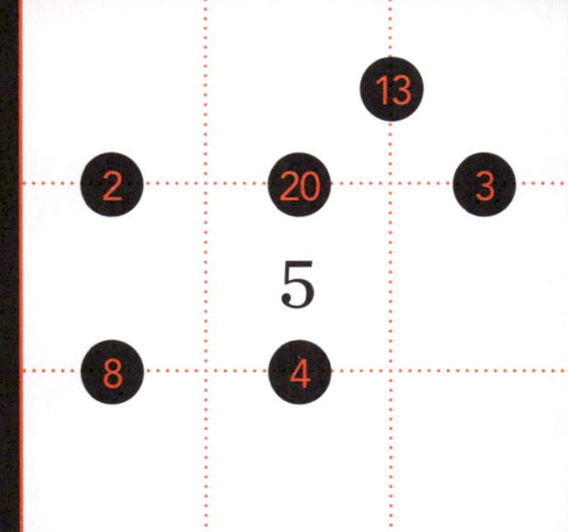

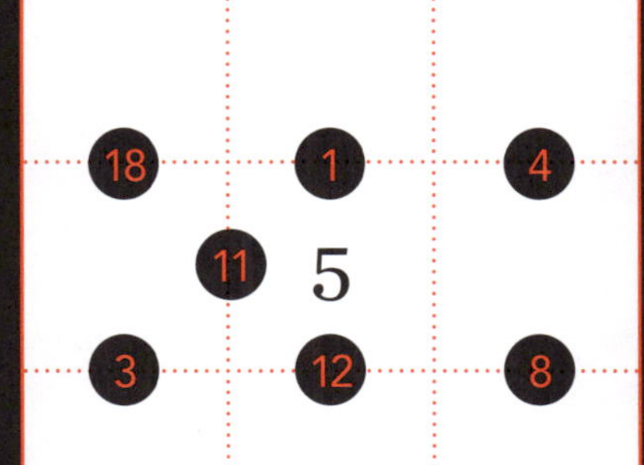

Enter the numbers 1 to 9 into each 3 × 3 grid so that each circled number is the result of adding, subtracting, multiplying, or dividing the two numbers in the cells it touches.

Head on back over to 99 to read more about these astonishing puzzles fresh from the mind of Sean Gardiner!

48

Waaaay back in '48

At number 96 in our countdown we met a formula involving the wondrous number pi expressed as an infinite sum.

Well, the hits just keep on coming! This one really fascinates me.

As with so many incredible observations about infinite series and pi, it was formulated by probably the greatest mathematician of them all, Leonhard Euler, back in 1748.

If you've read any of my books, or ever listened to me talk, or met me ... well, you probably know that I have a full-on geek-crush on Euler. What a legend.

He was born in Basel, Switzerland in 1707, making him 41 when he cracked this particular nut.

He died in Saint Petersburg in 1783.

We know from number 80 that the series $1 + 1/2 + 1/3 + 1/4 + 1/5 + \ldots$ keeps getting bigger forever, or, it 'diverges to infinity'.

But by subtracting instead of adding some of the terms we get this remarkable result:

$$\pi = 1 + 1/2 + 1/3 + 1/4 - 1/5 + 1/6 + 1/7 + 1/8 + 1/9 - 1/10 + 1/11 + 1/12 - 1/13 + \ldots$$

It might be difficult by looking at these first 13 terms to spot the pattern that determines whether the individual term in the sum attracts a plus or a minus sign.

Thankfully the great mathematical historian Carl Boyer explains it this way.

'After the first two terms, the signs are determined as

follows: if the number underneath the 1, we call this the "denominator", is a prime number of the form $4m - 1$ the sign is positive; if the denominator is a prime of the form $4m + 1$ then the sign is negative; for composite numbers the sign is equal to the product of the signs of the factors.'

Whoa, Carl! Settle down, big fella. What on Earth do you mean?

It's not actually that bad.

The prime number 11 is of the form $4 \times 3 - 1$ so the term $1/11$ attracts a + and 13 is of the form $4 \times 3 + 1$ so we have $-1/13$ in the series. For the term $1/10$, think of 10 as 2×5 and the $+1/2$ and $-1/5$ show us we will get the term $-1/10$.

Phew!

Before you go and make yourself a cup of coffee and sit and reflect on all this awesome knowledge, have a crack at this one.

Is the term $1/48$ positive or negative in the series?

How about $1/25$, $1/70$ and $1/2379$?

Get onto it!

48

You down with CRT?

During the year, I came across this this little quiz on news.com.au, where it was billed as the world's shortest IQ test.

The quiz, called a Cognitive Reflection Test or CRT, was developed at Princeton University in 2005 by psychologist Shane Frederick and, in its original form, contained just 3 questions.

The idea was to test your ability to override your 'gut reaction' and take your time thinking about the questions. Sometimes when an answer leaps out at you that might not be because you're brilliant (though if you're reading this book we can both safely assume that you are!). Maybe the question is well written to be misleading and trigger a quick, but wrong response.

Psychologists would more formally express this as 'can you ignore your intuition (system 1 type thinking) and let it give way to analysis (system 2 type thinking)'?

While it is written up online as a 60-second method to distinguish between Marie Curie and Homer Simpson, others suggest it correlates modestly but not precisely with IQ test intelligence.

To best get a taste for cognitive reflection, first answer the 3 questions in the sidebar as fast as you possibly can ... 5 seconds each *tops*.

Write down the first answer that leaps into your mind. Then go back, take as long as you like and see if you change your mind on any answers, or if your 'gut instinct' was correct.

Ready to race? Off you go! Answers at the back.

1. A bat and ball cost $1.10 in total. The bat costs $1.00 more than the ball. How much does the ball cost?

2. If it takes five machines five minutes to make five widgets, how long would it take 100 machines to make 100 widgets?

3. In a lake, there is a patch of lily pads. Every day, the patch doubles in size.

   If it takes 48 days for the patch to cover the entire lake, how long would it take for the patch to cover half the lake?

48

Jiggin' a jig

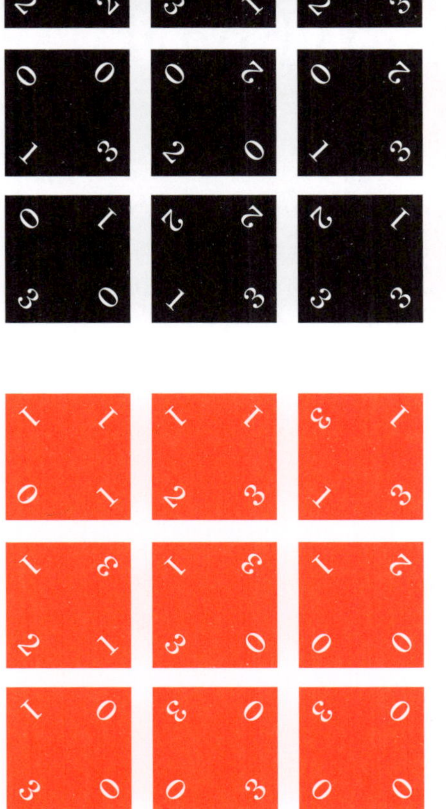

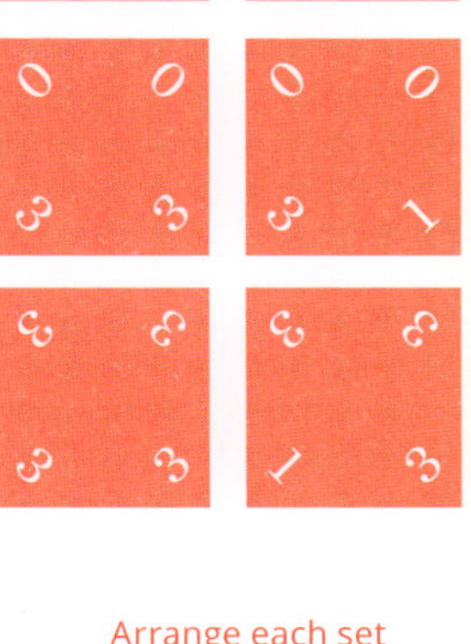

Arrange each set of 9 tiles into a 3 × 3 square so that adjacent tiles show the same numbers along their touching edges. Tiles may need to be rotated, but never reflected. Each set has only one solution. Check number 98 for more information.

For a downloadable cutout, head over to adamspencer.com.au/resources/

47

The way to my (clogged) heart

The brilliant Malcolm Gladwell was good enough to provide a quote for the cover of this book. So allow me to return the favour by giving a shout out to his wonderful podcast *Revisionist History,* which debunks myths we've come to take for granted.

In one of the very best episodes, 'McDonald's broke my heart', he sings a love song to the original McDonald's French Fry which he feels tasted perfect, until about 1990 when McDonald's changed the cooking oil.

The myth he debunks is that the new cooking process is actually *less* healthy than the old one, despite it being health concerns that brought about the change.

I'll leave the details for you to track down in the podcast, but the old beef tallow that was used to create these perfect fries was called Formula 47.

The name came from the price of McDonald's original 'All American Meal', which included a hamburger (15c) and fries (12c) and a 20c shake.

47

New high score!

Welcome back to the game of High Scoring Equation where you fill in the blanks to win ~~real cash prizes~~ fame and glory! Have fun!

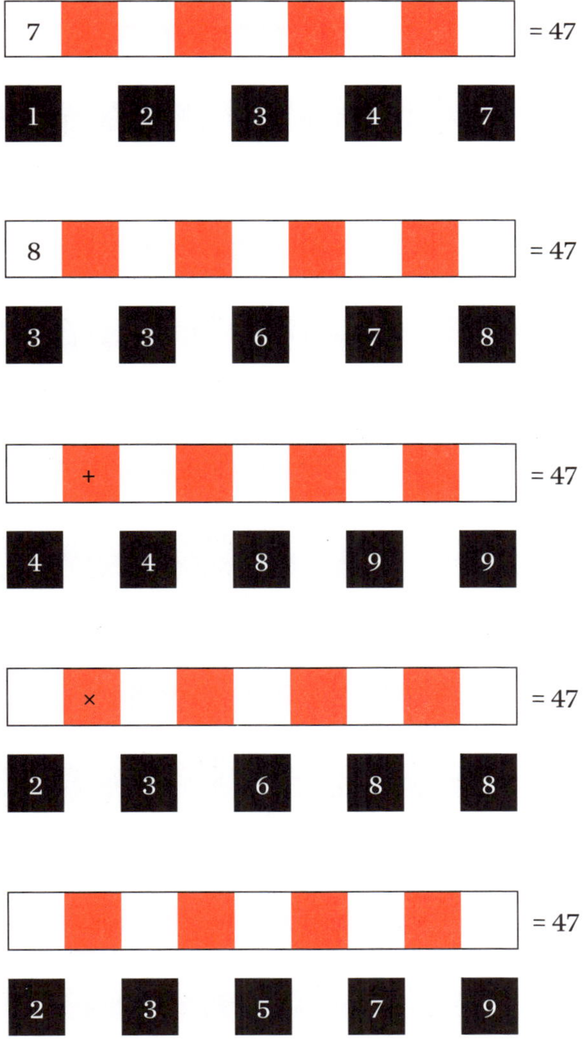

Reach the goal number by creating an equation using the provided numbers and your own choice of operators.

- The numbers should be placed in the white squares.
- Operations (+ – × ÷) are placed in orange squares.
- Order of operations matters, and you can't use brackets!
- Read the numbers off in order for your final score.
- Your goal is to find the equation that gives the highest score.

Head back over to number 97 if you need a refresher on these, otherwise ... get cracking!

47

Con tact

Victor Lustig's 10 rules of the con

1. Be a patient listener (it is this, not fast talking, that gets a con-man his coups).
2. Never look bored.
3. Wait for the other person to reveal any political opinions, then agree with them.
4. Let the other person reveal religious views, then have the same ones.
5. Hint at sex talk, but don't follow it up unless the other fellow shows a strong interest.
6. Never discuss illness, unless some special concern is shown.
7. Never pry into a person's personal circumstances (they'll tell you all eventually).
8. Never boast. Just let your importance be quietly obvious.
9. Never be untidy.
10. Never get drunk.

Can I say at the outset of this salute to the number 47 that conning people for a living is not cool. Ripping people off is not a good way to spend your life and can only lead to harm both to others and to your very sense of being.

But of all the people who have ever walked this Earth who would not care for my opinions on this subject, surely the best of the best (or perhaps the worst of the worst) was Victor Lustig.

'Count' Victor used anything up to 47 aliases during his life of conning. Imagine trying to keep track of which of the 47 people you sometimes pretended to be you were actually pretending to be at that given moment!

He carried dozens of fake passports, was wanted by over 40 American and international law agencies, spoke 5 languages, once swindled legendary gangster Al Capone out of $5000 and ... get this ... 'sold' the Eiffel Tower to an unsuspecting victim ... twice! (That's two different victims; not the same doofus two times.)

Having escaped one inescapable prison he was again captured and in 1935 sent to Alcatraz where he died 12 years later.

One way in which Lustig is immortalised is his famous 'Ten rules of the con'. I repeat them here not to serve as a guidebook for potential shysters and rip-off merchants, but to give an insight into the mind of the 20th century's greatest con artist.

A right-royal ripper of a riddle!

A great holiday tradition in the UK is the annual Royal Statistical Society Christmas Quiz which has melted new year brains for 24 years now.

The 2017 quiz was an absolute brute and among the 13 mind benders was this numerical nasty. Take it away, quiz-master Dr Tim Paulden!

'While browsing the mathematics section of his local second-hand bookstore, Stefan stumbles upon a dusty leather-bound volume. Curious, he opens the book and notices that it has suffered significant damage at the hands of a previous owner, with some of the text being obscured by black ink blotches. As Stefan leafs through the pages, the two lines above catch his eye – the second of which has fallen victim to two particularly unfortunate blotches.

'Always fond of a challenge, Stefan wonders whether it might be possible to figure out the values of the obscured numbers – assuming, of course, that both lines of numbers follow a consistent pattern. Can you help him out?'

As I said, this is a wickedly hard problem.

I've given a hint if you'd like it at the side. And the answer? Well, it's at the back of the book.

Start with the 2 on the left. First we do something with the other 4 numbers, (3, 11, 23 and 31); then we add 1; then we involve the 2. At the end of that we get the answer 11,765.

We repeat this process moving across number by number. So do something to (2, 11, 23, 31), add 1, involve 3 in some way and get 5229.

Keep following this process. Once you're sure you've got the rule right, what numbers must you need down the bottom so the rule works again?

46

All power to you

Check out these cool powers of 2 starting with 2^{46}

$2^{46} = 70368744177664$

$2^{56} = 72057594037927936$

$2^{66} = 73786976294838206464$

$2^{76} = 75557863725914323419136$

$2^{86} = 77371252455336267181195264$

$2^{96} = 79228162514264337593543950336$

$2^{106} = 81129638414606681695789005144064$

Is it just a fluke that the terms from 2^{46} up to 2^{96} all begin with a 7? What can you see going on here? Why did it stop for 2^{106}?

All POWER to you if you can figure it out. Ha! Geddit? Check your answer at the back ...

More hexes!

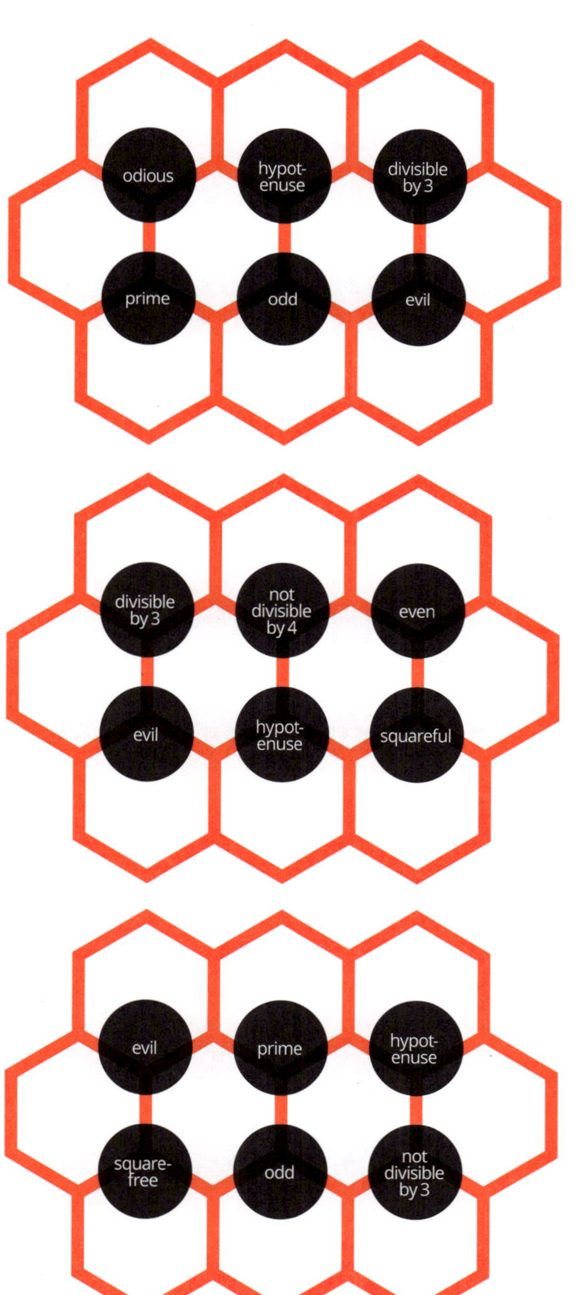

Place the numbers 41 to 50 into each hexagonal grid to satisfy all the rules.

Each cell is touching one or more categories in circles. The number in each cell must belong to all the categories touching it.

Need a hint? If the definitions are a bit tough for you, there's a full list of what these words mean at the end of the book.

Try-angular

Meet the triangle numbers T_n so called because ... well it should be obvious ... Take a look in the sidebar on the left.

Once you've seen the pattern you should be able to realise, without diagrams, that they continue $T_6 = 21$ followed by 28, 36, 45, 55, 66, 78 and so on.

Can you make sense of the general formula for triangular numbers, $T_{n+1} = T_n + (n + 1)$?

But look what happens when we add consecutive triangular numbers:

$T_8 + T_9 = 36 + 45 = 81 = 9^2$;
$T_9 + T_{10} = 45 + 55 = 100 = 10^2$.

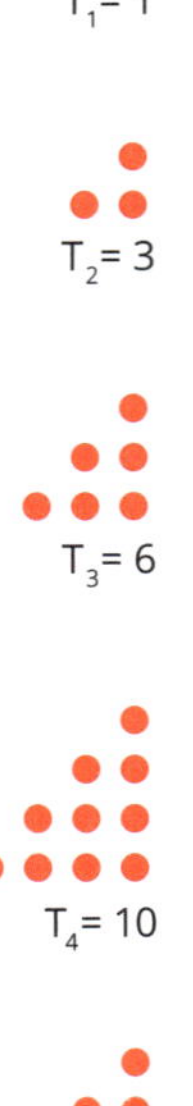

Can you see the pattern emerging? It looks like in general $T_{n-1} + T_n = n^2$. If that's not clear, look at other examples using consecutive triangular numbers.

Wanna see why — in a really cool way? Let's look at one of the simpler examples to see what's going on.

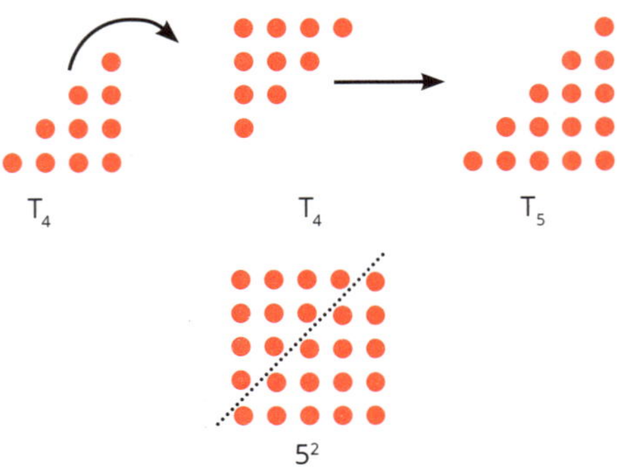

So there you go! The sum of consecutive triangular numbers is always a square.

On the tiles

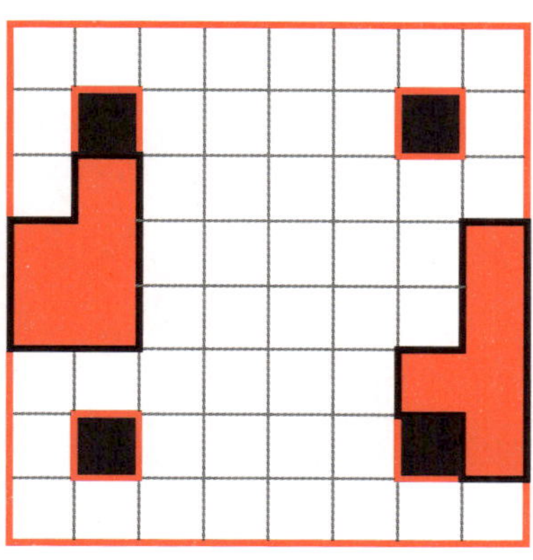

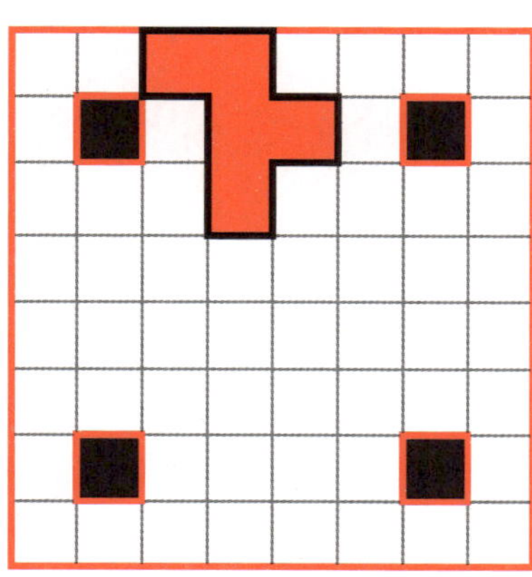

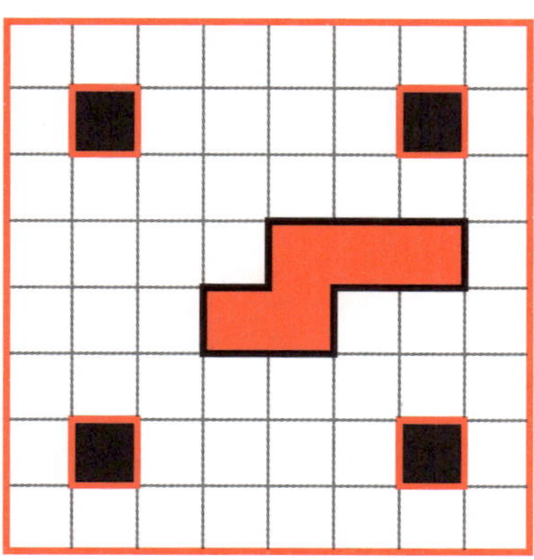

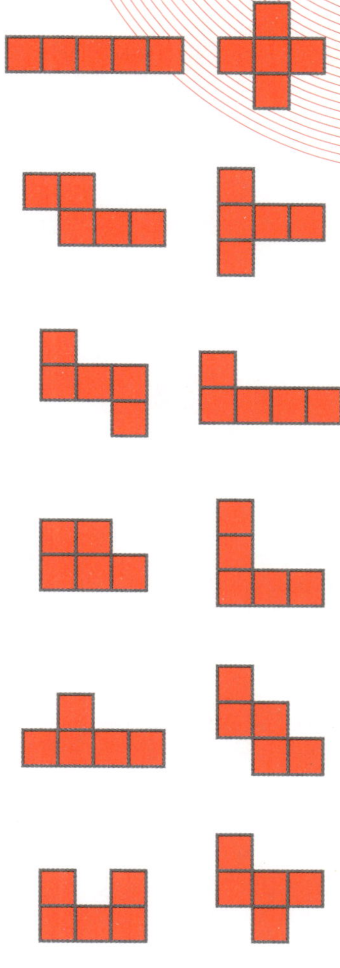

In each of these grids, finish tiling it so that it contains one of each pentomino type.

The 12 different pentominoes are shown here.

You may need to rotate and/or even flip some of them to get them to fit.

Head to adamspencer.com.au/resources if you'd like to download a copy of the puzzle to cut up!

44

Ruthenium sits at number 44 on the periodic table

In 2018 we discovered that atoms of ruthenium can be arranged in a certain way to make it magnetic at room temperature.

It is only the 4th element to have this quality, joining iron, cobalt and nickel.

Welcome to the club, ruthenium!

~~Dread~~gridlocks

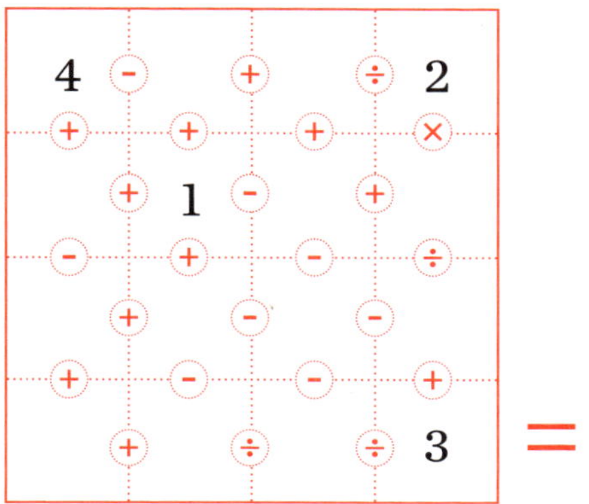

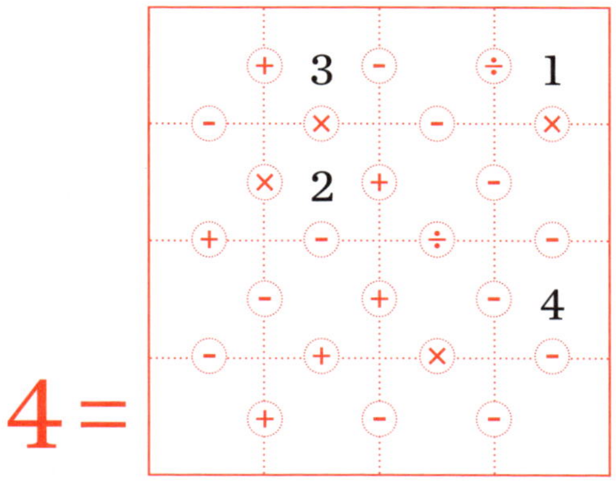

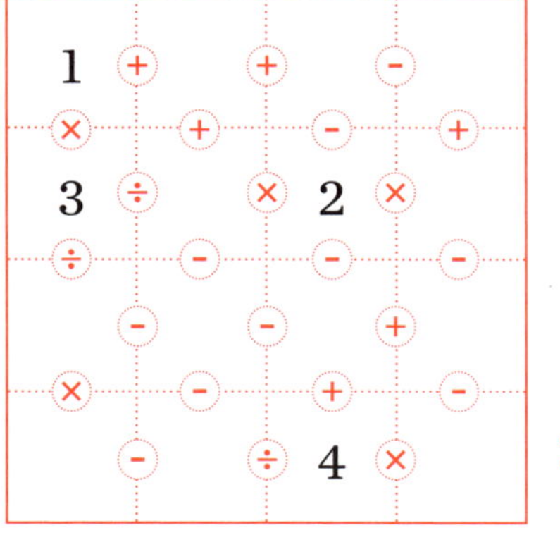

The rules for these are easy. The answers, well, they're a little tougher!

For each 4 × 4 grid, enter the numbers 1 to 16 so that each row and column equals the target number. As always, order of operations matters.

Can you break the gridlock?

If you don't like defacing this book, go to adamspencer.com/resources for a template to fill in.

43

Shine bright like Diophantine

We've clocked a few great hits in this countdown from the world of Diophantine equations back at 90, 84, 63 and 54 ... yeah I love them! And for those of you who've enjoyed the ride so far, here's another non-stop block of Diophantine rock.

In 1934 a mathematician named S. Sastry (sorry dude, couldn't find your first name anywhere)[1] wrote a paper titled 'On Sums of Powers'. In this modest sounding tome he produced the result $7^5 + 43^5 + 57^5 + 80^5 + 100^5 = 107^5$.

This in and of itself is awesome. But what really rocks is Sastry's observation that there are in fact an infinite number of results like this where a 5th power can be written as the sum of five 5th powers.

To explain what The Sass was on about, let's start with an easier example.

You may remember from the dim dark recesses of your mind, or from last week if you're in about Year 8, Pythagoras's Theorem which says that $a^2 + b^2 = c^2$ for the sides of right-angled triangles[2].

This has integer solutions like $3^2 + 4^2 = 5^2$ and $5^2 + 12^2 = 13^2$ and so we call the triples of numbers (3,4,5) and (5,12,13) Pythagorean triples.

There are infinitely many Pythagorean triples and we can see this if we know how to do a little bit of high school algebra. It turns out for any numbers x and y if we set $a = x^2 - y^2$, $b = 2xy$ and $c = x^2 + y^2$, then you will get: $a^2 + b^2 = c^2$.

For example, the case $x = 2$ and $y = 1$ gives $3^2 + 4^2 = 5^2$. Similarly $x = 3$ and $y = 2$ provides us with $5^2 + 12^2 = 13^2$.

You might like to plug a few more values of x and y into this process and see which Pythagorean triples you get.

[1] It's possible — although extremely unlikely, I grant you — that the letter stands for nothing at all.

US President Harry S. Truman's middle name was ... just the letter 'S'. When little Harry was born in 1884 his parents couldn't decide between his paternal grandfather (Shippe) and his maternal grandfather (Solomon).

So 'S' it was, and the rest is history.

[2] Remember this little number?

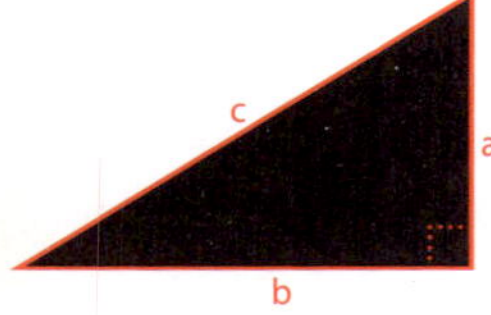

Ahem. I said 'you MIGHT LIKE TO plug a few more values ...' I've got all day, Year 8.

What Sastry did in 1934 was show another example of this sort of relationship which used two letters *u* and *v* (the names of the letters are of course irrelevant but as a shout out to The Sass I'm using the letters he did) to generate the equation:

$$(75\,v^5 - u^5)^5 + (u^5 + 25\,v^5)^5 + (u^5 - 25\,v^5)^5 + (10\,u^3\,v^2)^5 + (50\,u\,v^4)^5 = (u^5 + 75\,v^5)^5$$

In the same way that the infinity of possible integer values of x and y generated an infinite number of Pythagorean triples, so too here values of u and v give us an infinite list of 5th powers being written as the sum of five 5th powers.

So when $u = 2$ and $v = 1$ we get $(75 \times 1^5 - 2^5)^5 + (2^5 + 25 \times 1^5) + (2^5 - 25 \times 1^5)^5 + (10 \times 2^3 \times 1^2) + (50 \times 2 \times 1)^5 = (2^5 + 75)^5$ which takes us right back to the start of this block of fun with Sastry's example of $7^5 + 43^5 + 57^5 + 80^5 + 100^5 = 107^5$.

It's a little bit different here because for example taking $u = v = 1$ makes the term $u^5 - 25v^5 = 1^5 - 25 \times 1^5 = -24^5$. But if you add the -24^5 to each side once you're done you get an equation of the form $10^5 + 26^5 + 50^5 + 74^5 = 24^5 + 76^5$, showing that the number 2,543,488,000 can be written as both the sum of two and four 5th powers.

Experiment with small values of u and v and see what Diophantine equations you can generate.

I've listed a few at the back of the book for you.

I'm pretty sure I've convinced you that Diophantine equations shed light on some amazing equations and relationships between whole numbers.

But you may be wondering, 'yeah, sure, Adam ... but how do they affect my day-to-day life'?

Well, apart from giving you some great facts to drop at parties making you *the* person to be with, Diophantine equations (and related geometric constructions called elliptic curves) are now being used in cryptography to make your passport more secure.

There you go!

43

SADNESS The eyelids droop as the inner corners of the brows rise and, in extreme sadness, draw together. The corners of the lips pull down, and the lower lip may push up in a pout.

SURPRISE The upper eyelids and brows rise, and the jaw drops open.

ANGER Both the lower and upper eyelids tighten as the brows lower and draw together. Intense anger raises the upper eyelids as well. The jaw thrusts forward, the lips press together, and the lower lip may push up a little.

CONTEMPT This is the only expression that appears on just one side of the face: one half of the upper lip tightens upward.

DISGUST The nose wrinkles and the upper lip rises while the lower lip protrudes.

FEAR The eyes widen and the upper lids rise, as in surprise, but the brows draw together. The lips stretch horizontally.

HAPPINESS The corners of the mouth lift in a smile. As the eyelids tighten, the cheeks rise and the outside corners of the brows pull down.

Put a (fake) smile on yer dial

According to the world expert in reading facial cues and expressions, Paul Ekman, there are 43 muscles in the face that combine to create any expression you are making at any given time.

The 43 muscles have cool names like depressor supercilii, levator labii superioris, alaeque nasi and my favourite which you'll find in your cheek ... the buccinator!

Ekman, a professor of psychology at the University of California, has studied tens of thousands of people and his Facial Action Coding System (FACS) — the subtle clues to look for when someone is lying — teaches law enforcement authorities. He has also been a consultant to Hollywood studios to help create believable characters in animations.

He has even trained himself to move each of the 43 muscles in his own face individually so he can, on command, produce any emotional expression possible. This involves controlling 'microexpressions' that may last only half a second, and which we non-experts don't even realise we are doing.

Researchers around the world have shown that facial expressions remain the same from culture to culture and group into the follow 7 categories, as summarised in *Discover Magazine* in 2005, and listed here on the left.

I guess that that should be enough to put a smile on your dial?

Drill bits (day 6) ...

Up steps #43 to take today's drill, Gavin Robertson, the brilliant Australian off-spinner and handy lower-order batsman who wore 43 when playing one day cricket for the Aussies.

In which order did the players stand around the circle?

- #43 stood at the front of the circle.
- There was 1 person between #3 and #13.
- From #33's perspective, #73 was opposite them.
- #33 stood closer to #63 than to #93.
- #93, #23, and #33 stood together in some order.
- There were 2 people between #13 and #43.
- #53, #93, and #23 stood together in some order.
- From #43's perspective, #73 was on the right side of the circle.

Ten players stood in a circle for their daily drill.

They are numbered 3, 13, 23, 33, 43, 53, 63, 73, 83 and 93.

Each day a different player is the drill leader, which by delightful coincidence, corresponds to the chapter number and 'front position'.

Your task, coach, is to work out the order they stood in the circle each day given the list of clues provided by these fun-loving sporty puzzlers.

Game on!

42

Forty-*phew*!

We've already discussed the amazing numbers called factorials.

Remember that we get these by multiplying positive whole numbers by all the whole numbers less than them down to 1.

Well, 42! = 42 × 41 × 40 × 39 × 38 × 37 × 36 × 35 × 34 × 33 × 32 × 31 × 30 × 29 × 28 × 27 × 26 × 25 × 24 × 23 × 22 × 21 × 20 × 19 × 18 × 17 × 16 × 15 × 14 × 13 × 12 × 11 × 10 × 9 × 8 × 7 × 6 × 5 × 4 × 3 × 2 × 1 = 1.405×10^{51}.

And we estimate that there are about 10^{50} atoms in the Earth.

So let's pause to think about this.

A realtively small group of, say 42 – roughly the number of people waiting for a bus to the city in the morning (okay, a bus on a dream commute during school holidays ...) can line up to get on that bus ... more ways than there are atoms in the entire Earth!

Awesome.

Snakes alive!

→

8	×	9	+	5
×	7	×	6	+
8	×	6	–	3
×	8	–	4	=
2	×	2	=	42

→

7	×	8	+	2
–	7	×	4	×
4	–	8	+	2
×	3	×	7	=
5	+	4	=	42

→

9	+	5	+	5
×	6	+	9	–
1	+	2	–	8
+	7	–	1	=
7	–	3	=	42

In these grids, you can make paths from the top-left corner to the bottom-right, by moving between adjacent squares.

You can move up, down, left or right, but you can never visit the same square twice.

And, you guessed it, the paths must trace out the correct equation to reach the target number in the bottom right-hand corner.

For each grid, there are 3 different paths that all trace out correct equations.

Head on back to number 92 (if you haven't been there already) for a refresher if you need it!

42

Big in Texas

Forty-two, also known as Texas 42, is played with a standard set of double 6 dominoes and is often referred to as the 'national game of Texas'.

This might not make immediate sense because Texas is a state not a nation, but hey, we can roll with that.

The game is similar to card games like bridge and 500 where partners sit opposite each other and place bids as to how many 'tricks' they think they can win.

Instead of suits in cards (diamonds, spades, hearts, clubs), you might play a hand based on the dominoes with 3s, or blanks, or pairs and so on.

According to one historian, the game was created by two local boys who enjoyed playing cards but were discouraged from doing so by local religious types who thought card-playing was a sign of the devil. The boys reinvented a card like game with dominoes that did not carry the stigma of satan. Again this might not make immediate sense, because it is pretty much exactly a card game just with dominoes for cards, but hey, we can roll with that.

It's called Texas 42 because in each hand there are 42 points up for grabs. The 4 players split the 28 dominoes in the set evenly. Teams win one point for each trick plus bonus points if they pick up the dominoes where the total 'pips' (dots) are multiples of 5. The 4-1, 3-2 and 5-0 dominoes are worth 5 points and the 6-4 and 5-5 a massive 10 points each.

In certain cases the dealer may have to perform an act called 'riding the mule'. Why it is called that may not make immediate sense, but hey ...

Oh, brother!

Rudyard Kipling, author of *The Jungle Book* (and countless short stories and poems, including the perennially 'everybody's favourite' poem, 'If—') was awarded the Nobel Prize for Literature at 42 years of age.

Not only was he the first English-language writer to receive the gong, but also its youngest recipient to date.

Now, that's interesting, but I've got to be honest and say that this is really just a useful fact to let me include this fun puzzle at number 42.

The following passage is from Kipling's 1910 story 'Brother Square-Toes':

'I'll have to bide ashore and grow cabbages for a while, after I've run this cargo; but I do wish' — Dad says, going over the ***lugger's side with our New Year presents under his arm and young L'Estrange holding up the lantern — 'I just do wish that those folk which made war so easy had to run one cargo a month all this winter. It 'ud show 'em what honest work means.'***

'Well, I've warned ye,' says Uncle Aurette. 'I'll be slipping off now before your Revenue cutter comes. Give my love *to sister and take care o' the kegs. It's thicking to southward.'*

There's something noteworthy about the coloured, bold text here (aside from it being coloured and bold, smarty pants).

Can you figure out what?

Here's a hint, if you want it. I could also have included this puzzle under the number 66 ...

And a further hint (this puzzle is really hard!) look at the last word before and after the coloured text.

41

The maths of the paaaaaaaarty

Five girls are planning a party.

They have decided to invite some boys to the party — ew, yuck — some of whom they know and some who are complete strangers.

For reasons that remain vague, they want to be certain that among the group of invited boys, there is a set of 3 boys such that 3 of the 5 girls will either be friends with all 3 boys or strangers to all 3 boys.

What's the minimum number of boys they need to invite to make sure this strange desire is fulfilled?

Okay, you might have struggled with this, but you're probably bright enough to look at the number this fact is provided under and guess, 'Um, Adam ... is it 41?'

Indeed it is, Sherlock. That's right, no matter which of the boys knows or doesn't know any of the girls, if they invite *any* group of 41 boys, the girls will have satisfied the criteria set above.

It's actually not *too* hard to deduce. The worst case scenario is that for every triple of girls, exactly 2 boys know only them and exactly 2 boys don't know them (and do know the other 2 girls). So for each triple of girls, 4 boys can exist without satisfying the condition. There are 10 ways to pick the 3 girls from the group of 5 (see the sidebar as to why), so in the worst case scenario, 40 boys can be invited without the condition being satisfied. Which means ... 41 is the minimum amount of boys that need to be invited.

While this might seem to be nothing more than an exercise in bizarre party planning preferences, this sort of question is actually related to a fascinating branch of mathematics known as graph theory, which we met back at number 55 and will visit again at number 10 with reference to one of the most famous maths movies of all time.

Call the girls A B C D E (without wishing to seem impersonal!) In choosing a group of 3 girls, there are 5 girls you could pick first, then 4 girls who could be second and 3 possible choices for the third girl.

So you could select 5 × 4 × 3 = 60 such groups of 3 girls.

But for every group of 3 girls so chosen, say A D E, the 6 groups ADE, AED, DAE, DEA, EAD and EDA are effectively the same group.

So we divide our big set of 60 groups by 6 to get 10 unique groups of 3 girls chosen from the 5.

The mathematicians among us say '5 choose 3 equals 10'.

Doubly-true (still!) alphametics

Okay, thrill-seekers, this is it.

Now, some of you have already been braving the world of doubly-true alphametics without the hints I've provided for a while already.

For the rest of you it's bungee-jumping time. No hint — no safety net. Again, the letters correspond to a unique digit each and the equation holds in word and numerical form.

Don't forget the lead digits on each line, the F, T and S here cannot equal 0.

Here goes ...

```
  FOURTEEN
       TEN
       TEN
+    SEVEN
----------
  FORTYONE
```

Have a crack at this little beauty in honour of good old number 41.

Hey, I said no hints, okay? Good luck! :)

41

PDF

The prodigious 17th century mathematician Pierre de Fermat is most famous for his 'Last Theorem'.

This theorem says that while there is an infinite number of examples like $3^2 + 4^2 = 5^2$, $5^2 + 12^2 = 13^2$ and so on, where a square is written as the sum of two squares, you will never get a cube that can be written as the sum of two other cubes, nor a 7th power that can be written as the sum of two other 7th powers, nor any case using powers higher than 2.

But this wasn't the only contribution that PDF (the dude, not the file type) made to our knowledge.

Fermat's Theorem on the sums of two squares says that you can write a prime number p as the sum of two perfect squares $x^2 + y^2$ if and only if you can write $p = 4n + 1$ where n is a whole number.

If you haven't seen the phrase 'if and only if' before, it is like an arrow that points both ways. The two requirements here, being the sum of two perfect squares and being of the form $4n + 1$, are effectively equivalent.

Now, 41 can be written in the form $41 = 4 \times 10 + 1$, so it can be written as the sum of two squares: $41 = 5^2 + 4^2$.

Question time! Find all the primes p under 100 that can be written in the form $p = 4n + 1$, and also write each of these primes as the sum of two squares.

PDF — the file format, not the awesome maths genius — was released by Adobe way back in 1993.

In 2015, Adobe's VP of Engineering for Document Cloud, Phil Ydens, estimated there may be up to 2.5 *trillion* PDF documents in the world ... and rising!

Adobe holds patents to PDF (Portable Document Format), but licenses them for royalty-free use in developing software complying with its specifications.

Thanks, Adobe!

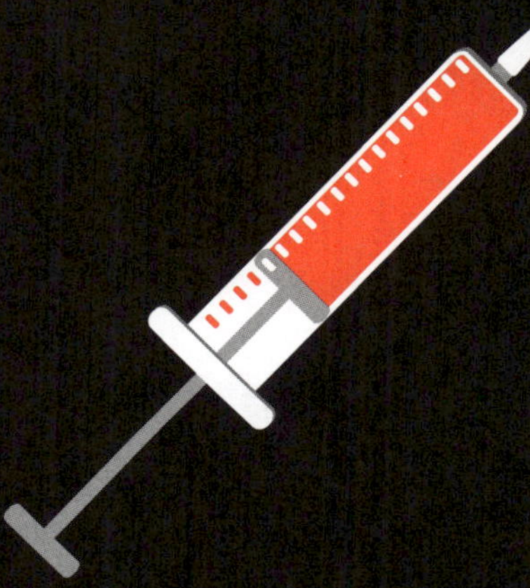

40% of Australians are blood type O-positive

So says the Australian Red Cross, which has pretty good intel on such matters.

The distribution of blood types overall is:

- O-positive: 40%
- O-negative: 9%
- A-positive: 31%
- A-negative: 7%
- B-positive: 8%
- B-negative: 2%
- AB-positive: 2%
- AB-negative: 1%

40 is the highest number ever counted to on *Sesame Street*

Although the number was first introduced to the Street in a Guy Smiley game show entitled 'Happiness Begins at 40' in 1986, it was many years before an epsiode was 'brought to you' by the number 40 (and the letter 'H', for the record).

That honour wasn't bestowed until the 40th anniversary show — episode 4187 — in 2009.

And while Big Bird rocks, Oscar the Grouch has a certain nihilistic charm about him and Mr Snuffleupagus is 'ridic' trippy, my favourite character on Sesame Street is obviously ... Count von Count.

Count von Count once told an interviewer that his favourite number was 34,969.

His reason?

$34{,}969 = 187^2$.

Legend.

Blankety blanks (#40)

Head on back to number 100 if you need a reminder of how these puzzles work. Because now you'll encounter your first BB with *no* clues! Rules on the right, pens at the ready ... aaaand go!

$$8 \times (\bigcirc + (\bigcirc - (\bigcirc - \bigcirc) \div \bigcirc) \div (\bigcirc - \bigcirc)) = 40$$

$$(\bigcirc + \bigcirc) \times ((\bigcirc - \bigcirc) \times (\bigcirc - \bigcirc \div 2) - \bigcirc) = 40$$

$$(\bigcirc - \bigcirc) \times ((\bigcirc + (\bigcirc + \bigcirc) \div \bigcirc) \div 3 + \bigcirc) = 40$$

$$(\bigcirc - (\bigcirc - \bigcirc) \div (\bigcirc \div \bigcirc + \bigcirc)) \times (7 - \bigcirc) = 40$$

and now ... *drumroll* ... you're on your own! Good luck!

$$(\bigcirc \times ((\bigcirc + \bigcirc) \div (\bigcirc + \bigcirc) + \bigcirc) - \bigcirc) \times \bigcirc = 40$$

Reach the target number by filling in the blanks in each equation.

A completed equation must incorporate each and every digit from 0 to 9.

You cannot move the operations (+, −, ×, ÷) found between blanks.

Order of operations applies!

39

Aussies throw away up to $39,000,000 of loose change every year!

Yep, according to a survey released by the bank ING in 2017, Australians really don't like loose change.

One in four people is irritated by carrying loose change in their pocket and one in nine of us just throws away the loose coins we receive!

Seriously, I'll happily come around and collect it all — just call me!

Sean's Syndesis

2 24
3
5 27 3
14

7 13
7
9 28 3
32

12
3
4 10
1 3
35

Enter the numbers 1 to 9 into each 3 × 3 grid so that each circled number is the result of adding, subtracting, multiplying, or dividing the two numbers in the cells it touches.

Head on back over to number 99 to read more about these cool puzzles fresh from the mind of Sean Gardiner!

Matvei Petrovich Bronstein

was a brilliant Soviet physicist and one of the original greats of quantum gravity; our attempts to understand how gravity works on the incredible small or quantum scale.

He wrote extensively on astrophysics, semiconductors, cosmology and quantum electrodynamics. He also wrote many science books for children.

In 1938, he was put to death without trial during a government campaign of executions referred to now as 'The Great Purge'.

Around 1,000,000 Russians were executed by their own government including Bronstein who was just 31. One can only imagine the discoveries he would have made across the rest of his life.

A senseless waste.

38

Alert *and* alarmed ...

On 13 January 2018 at 8.07 am the following text message appeared on the phones across Hawaii and ran as a banner on television channels.

> **EMERGENCY ALERT**
> BALLISTIC MISSILE THREAT INBOUND TO HAWAII. SEEK IMMEDIATE SHELTER. THIS IS NOT A DRILL.

For a state that had recently begun retesting warning sirens in response to rumours that North Korea could reach Hawaii with missiles, this understandably freaked more than a few people out.

It was not for another 38 minutes that a second message went out from the official agency describing the first one as a false alarm.

The Governor of Hawaii was alerted only 2 minutes after the initial message. He took another 15 minutes before announcing that on Twitter.

It turns out he'd forgotten his Twitter password!

38% of all cars on the road in 1900 ...

were estimated to have been ... electric.

38

Jiggin’ a gigue

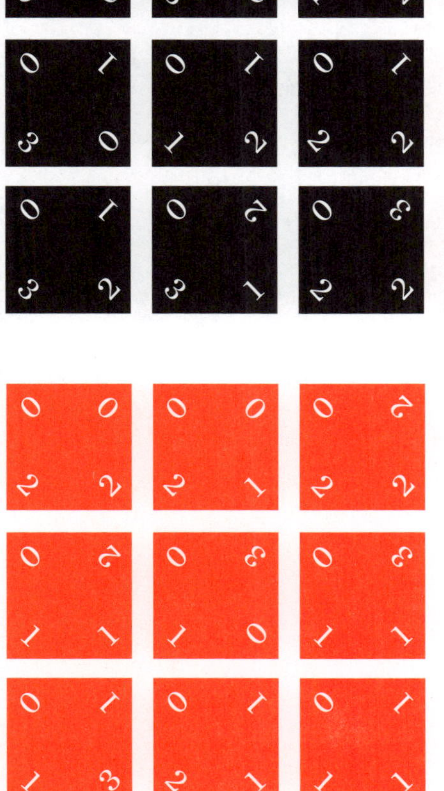

Arrange each set of 9 tiles into a 3 × 3 square so that adjacent tiles show the same numbers along their touching edges. Tiles may need to be rotated, but never reflected. Each set has only one solution. Check number 98 for more information.

For a downloadable cutout, head over to adamspencer.com.au/resources/

37

The good trivia buffs ...

at @qikipedia inform me that despite it being one of Britain's most closely-guarded properties, during Tony Blair's term as Prime Minister, 37 computers were stolen from 10 Downing Street (along with 4 mobile phones, 4 printers, 2 cameras, a video recorder, and a bicycle)!

New high score!

Welcome back to the game of High Scoring Equation where you fill in the blanks to win ~~real cash prizes~~ fame and glory! Only 3 hints this time. Have fun!

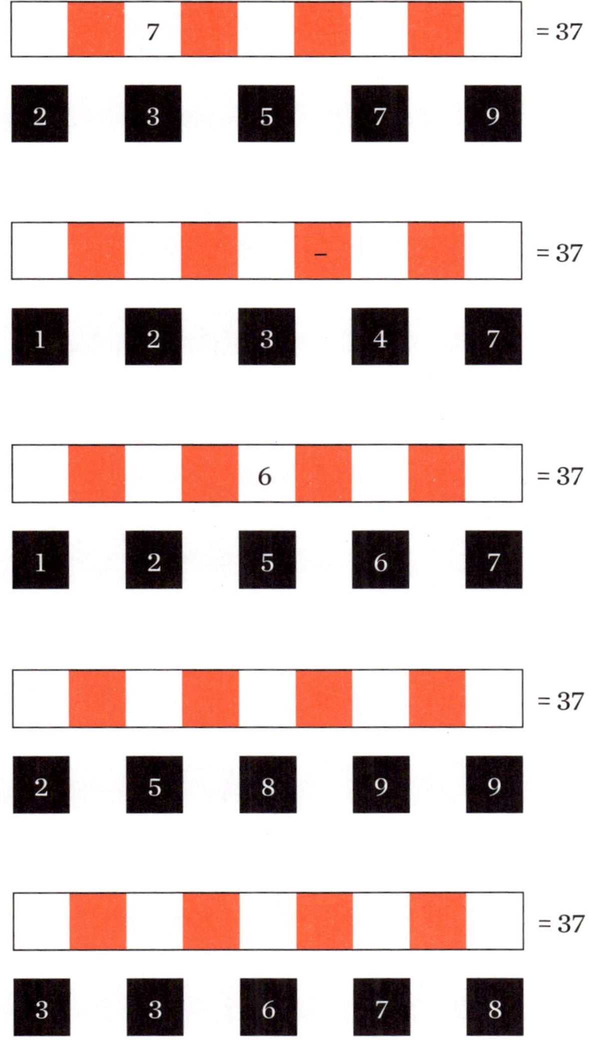

Reach the goal number by creating an equation using the provided numbers and your own choice of operators.

- The numbers should be placed in the white squares.
- Operations (+ – × ÷) are placed in orange squares.
- Order of operations matters, and you can't use brackets!
- Read the numbers off in order for your final score.
- Your goal is to find the equation that gives the highest score.

Head back over to number 97 if you need a refresher on these, otherwise ... get cracking!

36

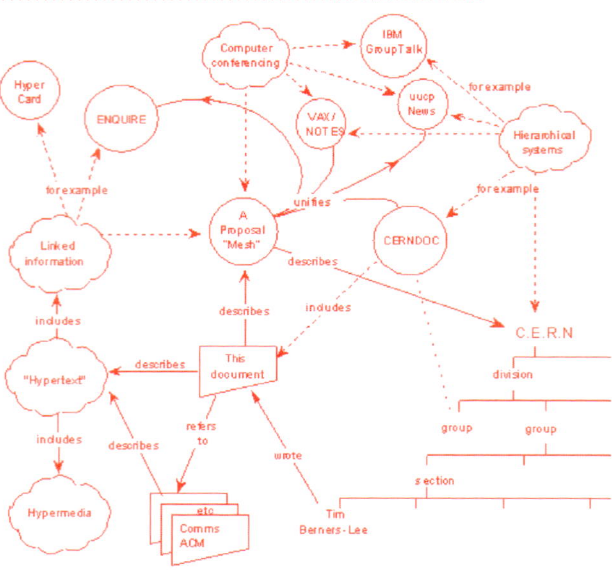

36 wwwords

You like to check your Insta a few times a day? Read the news online? Check out the odd clip on YouTube? Learn a new fact about algebra and number theory?* Well, it's all thanks to these 36 words:

This proposal concerns the management of general information about accelerators and experiments at CERN. It discusses the problems of loss of information about complex evolving systems and derives a solution based on a distributed hypertext system.

In 1989, with these two sentences, Tim Berners-Lee proposed an information management system for CERN (the European Organization for Nuclear Research) to make collaboration easier and to avoid loss of information in putting together complex documents across the agency's many arms.

At the time he referred to it as 'Mesh' but when writing the code in 1990, settled upon the name 'world wide web'.

And the rest, as they say ...

* Okay, maybe that's just me ...

More hexes!

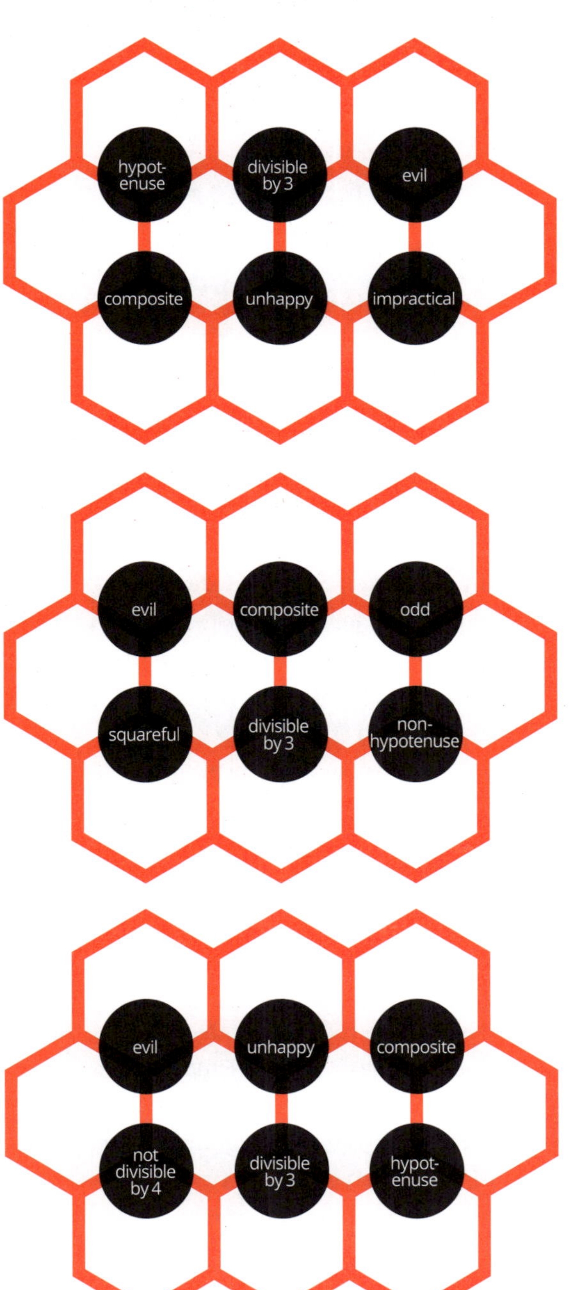

Place the numbers 31 to 40 into each hexagonal grid to satisfy all the rules.

Each cell is touching one or more categories in circles. The number in each cell must belong to all the categories touching it.

Need a hint? If the definitions are a bit tough for you, there's a full list of what these words mean at the end of the book.

35

A ripper of a right-royal researcher

The Royal Society is the oldest independent scientific academy in the world.

Founded in London in 1660, it is 'dedicated to promoting excellence in science' and being made a fellow is about as prestigious a gig as a scientist can hope for in their lives (with perhaps the exception of winning a Nobel Prize, or in the case of mathematics, a Fields Medal).

The youngest member of the Royal Society is Prince William who, when I wrote this, was 35. By the time you're reading this he'll be 36, but I already had a cracker of a story for that number up my sleeve.

Anyway, no offence to the Prince, his membership ain't *that* impressive because he is automatically a member, given the whole 'Royal' bit in the name.

But being the *second* youngest member of the Royal Society and the youngest *elected* Fellow? Now that's impressive.

Take a bow professor Geordie Williamson, an Australian mathematician from the University of Sydney who, at just one year older than Prince William, was elected to the Society in 2018.

Aussie, Aussie, Aussie!

On the tiles

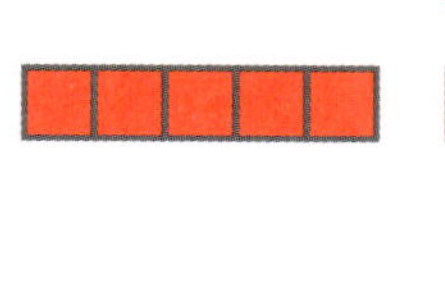

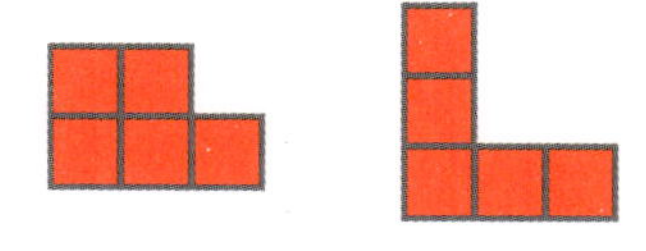

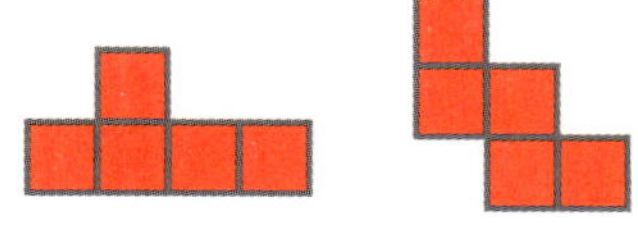

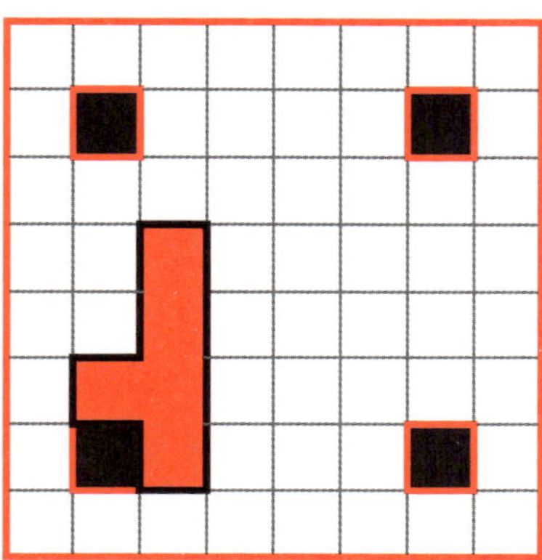

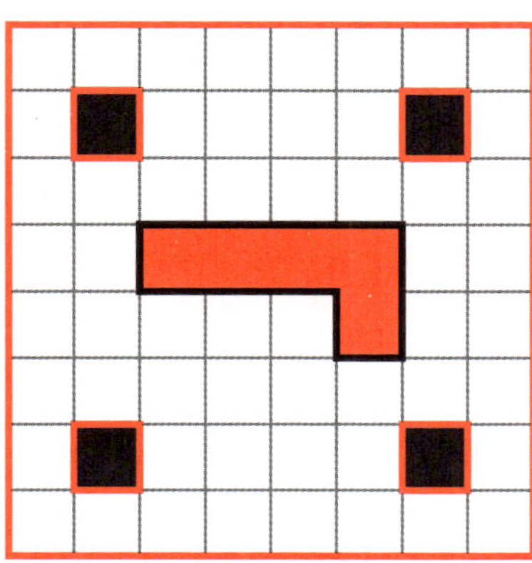

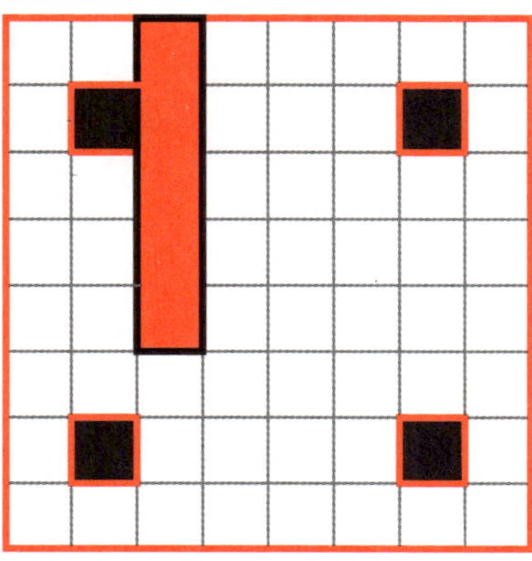

In each of these grids, finish tiling it so that it contains one of each pentomino type.

The 12 different pentominoes are shown here.

You may need to rotate and/or even flip some of them to get them to fit.

Head to adamspencer.com.au/resources if you'd like to download a copy of the puzzle to cut up!

34

Fibbin' to 100

We met the Fibonacci sequence back at 61 in our countdown. It is one of the most famous lists of numbers in all mathematics.

You might remember that 34 is indeed a Fibonacci number.

Well, here's something you might not know. We call two numbers 'relatively prime' if they have no common factor apart from 1.

So while neither 12 nor 25 are prime numbers, they are relatively prime because $12 = 2 \times 2 \times 3$ and $25 = 5 \times 5$ from which we can see they have no common factors.

A cool fact that relates prime numbers and the Fibonacci sequence is this:

> If m and n are relatively prime, then so are the Fibonacci numbers F_m and F_n.

Notice this is the converse statement to (an extension of) the one mentioned at number 61. Together they essentially say F_m and F_n share a factor greater than 1 if, and only if, m and n share a factor greater than 2.

Write out the Fibonacci numbers up to 100 and convince yourself that the rule holds for those numbers.

Still gridlocked!

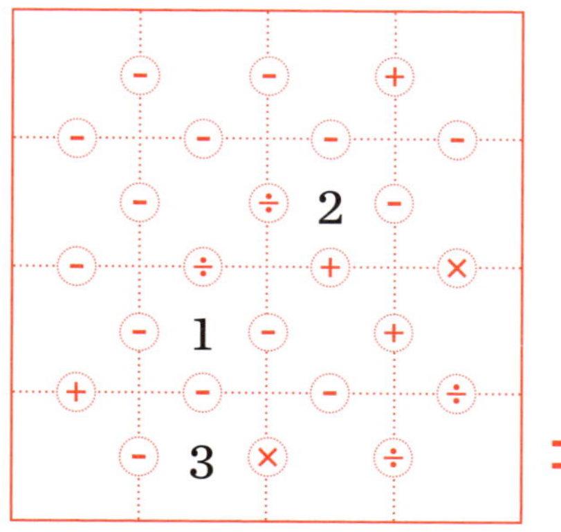

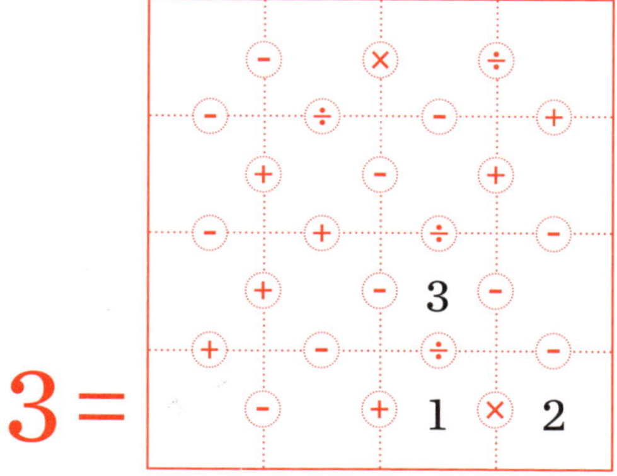

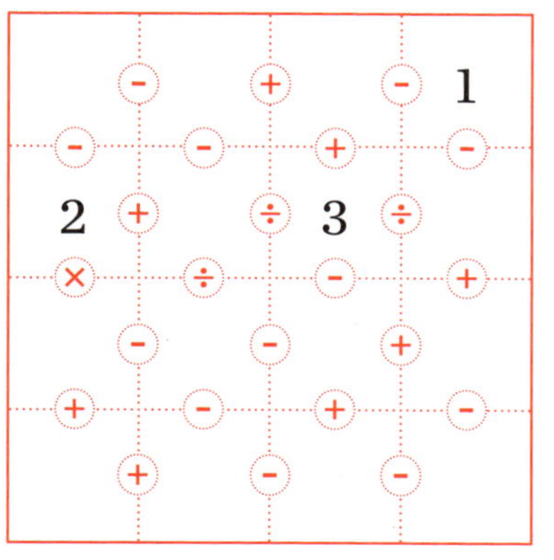

The rules for these are easy. The answers, well, they're a little tougher!

For each 4 × 4 grid, enter the numbers 1 to 16 so that each row and column equals the target number. As always, order of operations matters.

Can you break the gridlock?

33

Merlot grapes are produced in 33 countries around the world

So says the Australian wine-lovers bible *Halliday*.

This is more than shiraz (27 countries) but not as much as cabernet sauvignon which, being produced in 40 countries, is described as 'the world's most noble red variety, taking into account its quality and quantity'.

In Australia, the order is reversed. In 2017 Australia produced 245,000 tonnes of cabernet worth $178 million compared to 440,000 tonnes of shiraz worth $378 million.

Cheers to that!

Nerd props

Sincere nerd props to French puzzle designer Grégoire Pfennig of *Greg's Puzzles* who, in late 2017, produced a new world-record-sized Rubik's Cube.

Rather than the traditional 3 × 3 × 3 cube which most of us have seen and, if you're reading my book, I assume loved, or the 4 × 4 × 4 which is pretty easy to grab in a games shop, Grégoire 3D printed a gigantic 33 × 33 × 33 cube!

With 6153 parts, it is the most parts-heavy puzzle of any type ever made. It takes over 200 hours to build – much longer to solve I'm sure – and can be yours for as little as $23,000. So long as you're willing to wait about 3 months for him to belt one out for you.

While we're on the subject, another big shout out to Aussie cube-solving wunderkind Feliks Zemdegs who, in 2018, at the Cube for Cambodia event, set a new world record for the traditional 3 × 3 × 3 cube of 4.22 seconds*.

It's the 9th time Feliks has held the world record alongside his record for best average time over 5 solves, not to mention his record for solving a Rubik's Cube one-handed in ... wait for it ... 6.88 seconds.

Legend!

* YES! This is not a typo (and how dare you imagine there *cough* be any typos in my book. 4.22 seconds!

33

~~50~~ 33 ways to leave your lover

According to Shu Xin, a Chinese 'love detective' featured on a BBC Radio documentary in December 2017, there are 33 ways to dispel a mistress.

These include persuading the unfaithful person's boss to move them to another city, making the mistress fall in love with someone else and 31 other tricks in the love detective's tool box.

While we may giggle, it turns out people pay Shu Xin and the Weiqing Love Hospital, the most popular mistress-dispelling service in Shanghai, thousands of dollars for their services.

Drill bits (day 7) ...

Today the skills session for our first ever assembled international, cross code, mega-star sports team is run by the recently retired Hawthorn AFL great, the incredibly skilful #33 Cyril Rioli.

In which order did the players stand around the circle?

- #33 stood at the front of the circle.
- #63, #83, and #13 stood together in some order.
- From #83's perspective, #33 was opposite them.
- There was 1 person between #33 and #93.
- #73, #83, and #13 stood together in some order.
- From #93's perspective, #43 was opposite them.
- #53 stood closer to #3 than to #23.
- #53, #63, and #93 stood together in some order.
- From #83's perspective, #73 was on the right side of the circle.

Ten players stood in a circle for their daily drill.

They are numbered 3, 13, 23, 33, 43, 53, 63, 73, 83 and 93.

Each day a different player is the drill leader, which by delightful coincidence, corresponds to the chapter number and 'front position'.

Your task, coach, is to work out the order they stood in the circle each day given the list of clues provided by these fun-loving sporty puzzlers.

Game on!

32

Just two timpani and a microphone

Back in the day you only needed two timpani (those big cool looking drums to the back of an orchestra that musicians play standing up) to get the job done.

They were usually 29 inches tall and 26 inches across.

But a percussionist's gotta do what a percussionist's gotta do and these days, any self-respecting drum-thumper will be seen with 4 or even 5 timpani.

The widest timpano (the singular of timpani) is 32 inches across, and a typical set includes 29-, 26- and 23-inch drums.

Who knows, if you're a real playa, you just might have a cheeky 21-inch timp in your arsenal.

The word 'timpani' derives from the Latin word 'tympanum', meaning 'a hand drum'.

It's the little things Fer-matter

In 1732 the great Leonhard Euler showed that 4,294,967,297 = 641 × 6,700,417.

This proved many things. First, that Euler was a genius – but we didn't need any more proof of that by then.

He is today regarded as probably the greatest mathematician ever ... that's even including Matt Damon's character in *Good Will Hunting*, who we will meet a little later.

But it also disproved something that another great maths-mind had suggested years earlier.

Pierre de Fermat examined numbers of the form $2^{2^n} + 1$ (as we also saw back at number 61).

$$F_n = 2^{2^n} + 1$$

For the first few integer values of n we get (noting we start from the top of the powers and work down):

$$F_1 = 2^{2^1} + 1 = 2^2 + 1 = 4 + 1 = 5$$
$$F_2 = 2^{2^2} + 1 = 2^4 + 1 = 16 + 1 = 17$$
$$F_3 = 2^{2^3} + 1 = 2^8 + 1 = 256 + 1 = 257$$
$$F_4 = 2^{2^4} + 1 = 2^{16} + 1 = 65{,}536 + 1 = 65{,}537$$

... and all these values of F_n are prime numbers.

Pierre suggested all Fermat numbers would be prime. Euler showed F5 was not. In fact F4 is the largest known prime Fermat number, which happens to be – you guessed it – 4,294,967,297.

We have tested all the way up to F_{32} which is over a billion digits long!

32

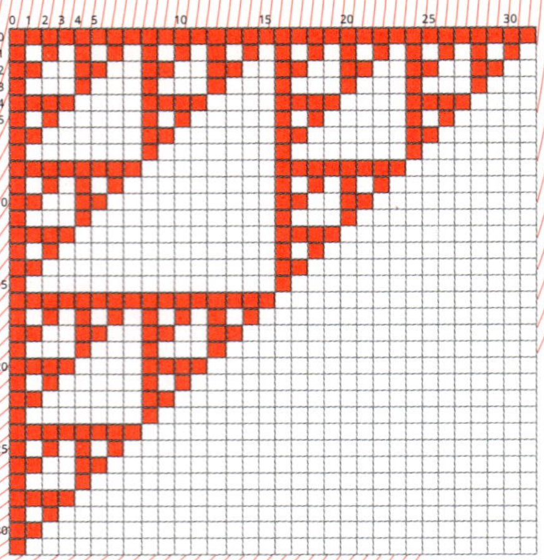

Touching base

We spoke earlier at number 49 about counting in different bases.

Base 2 is the number system that, among other things, is used to underpin computer coding and the amazing digital world in which we live. Base 2 arithmetic uses only 0s and 1s so to count from 0 up to what we think of as 10 we would count 0, 1, 10, 11, 100, 101, 110, 111, 1000, 1001, 1010 ...

Each place in a number represents a higher power of 2, so the number 10011101010 in binary represents:

$$1 \times 2^{10} + 1 \times 2^{7} + 1 \times 2^{6} + 1 \times 2^{5} + 1 \times 2^{3} + 1 \times 2^{1} = 1024 + 128 + 64 + 32 + 8 + 2 = 1258.$$

Cruising around Twitter one day I spotted this beauty from the good people at the *Republic of Math*. Let's take all 32 of the integers from 0 up to 31 and write them in base 2.

Now compare them. If two numbers a and b do not have a 1 in a common position in their binary expression, colour in the square positioned at (a,b) on a grid. If they have a common 1, leave the square blank.

In binary, 4 = 100, 6 = 110 and 17 is 10001. So spots (4,6) and (6,4) would be coloured in, but (4,17) (17,4), (6,17) and (17,6) would be left blank.

You get the gorgeous pattern above*.

* Bonus fact: which is also what you get if you colour in all the odd numbers in Pascal's triangle ...

Snakes alive!

→

5	–	5	×	7
×	3	+	6	×
6	+	1	+	3
×	8	×	8	=
1	×	9	=	32

→

9	×	9	–	6
–	9	–	5	+
7	–	1	×	2
–	3	×	3	=
1	×	2	=	32

→

5	–	8	–	2
×	3	×	6	–
6	×	7	×	1
×	2	+	6	=
7	×	6	=	32

In these grids, you can make paths from the top-left corner to the bottom-right, by moving between adjacent squares.

You can move up, down, left or right, but you can never visit the same square twice.

And, you guessed it, the paths must trace out the correct equation to reach the target number in the bottom right-hand corner.

For each grid, there are 3 different paths that all trace out correct equations.

Head on back to number 92 (if you haven't been there already) for a refresher if you need it!

31

XUKEXWSLZJUAXUNKIGWFSOZRAWURORKXAOS
LHROBXBTKCMUWDVPTFBLMKEFVWMUXTVTWUI
DDJVZKBRMCWOIWYDXMLUFPVSHAGSVWUFWOR
CWUIDUJCNVTTBERTUNOJUZHVTWKORSVRZSV
VFSQXOCMUWPYTRLGBMCYPOJCLRIYTVFCCMU
WUFPOXCNMCIWMSKPXEDLYIQKDJWIWCJUMVR
CJUMVRKXWURKPSEEIWZVXULEIOETOOFWKBI
UXPXUGOWLFPWUSCH

WURVFXGJYTHEIZXSQXOBGSVRUDOOJXATBKT
ARVIXPYTMYABMVUFXPXKUJVPLSDVTGNGOSI
GLWURPKFCVGELLRNNGLPYTFVTPXAJOSCWRO
DORWNWSICLFKEMOTGJYCRRAOJVNTODVMNSQ
IVICRBICRUDCSKXYPDMDROJUZICRVFWXIFP
XIVVIEPYTDOIAVRBOOXWRAKPSZXTZKVROSW
CRCFVEESOLWKTOBXAUXVB

Feeling Feyn, man

In 1987, Chris Cole posted a message on the sci.crypt Usenet (sigh ... ask your parents, kids ...):

When I was a graduate student at Caltech, Professor Feynman showed me three samples of code that he had been challenged with by a fellow scientist at Los Alamos and which he had not been able to crack. I also was unable to crack them. I now post them for the net to give it a try.

Note that the Professor in question is the legendary Richard Feynman who, to over-summarise his acheivements for the sake of space, was a Nobel laureate, so you're right to imagine they're pretty tough.

Well, maybe not *too* tough, in the first case. Labelled 'Easier', it was cracked the next day by Jack Morrison of NASA's JPL. He wrote, 'It's a pretty standard transposition: split the text into 5-column pieces, then read from lower right upward ...' And the other two?

Well, they're harder. In fact, they're still unsolved, over 30 years later. Have a look above, if you're keen, and be sure to let me know if you crack them!

MEOTAIHSIBRTEWDGLGKNLANEA
INOEEPEYSTNPEUOOEHRONLTIR
OSDHEOTNPHGAAETOHSZOTTENT
KEPADLYPHEODOWCFORRRNLCUE
EEEOPGMRLHNNDFTOENEALKEHH
EATTHNMESCNSHIRAETDAHLHEM
TETRFSWEDOEOENEGFHETAEDGH
RLNNGOAAEOCMTURRSLTDIDORE
HNHEHNAYVTIERHEENECTRNVIO
UOEHOTRNWSAYIFSNSHOEMRTRR
EUAUUHOHOOHCDCHTEEISEVRLS
KLIHIIAPCHRHSIHPSNWTOIISI
SHHNWEMTIEYAFELNRENLEERYI
PHBEROTEVPHNTYATIERTIHEEA
WTWVHTASETHHSDNGEIEAYNHHH
NNHTW

By all means have a crack at this 'Easier' one if you'd like. The answer (the first 8 lines or so of Chaucer's *Canterbury Tales*) is below, though I'm afraid you'll have to squint ... check it out online if you're keen:

WHANTHATAPRILLEWITHHISSHOURESSOOTET
HEDROGHTEOFMARCHHATHPERCEDTOTHEROOT
EANDBATHEDEVERYVEYNEINSWICHLICOUROF
WHICHVERTUENGENDREDISTHEFLOURWHANZE
PHIRUSEEKWITHHISSWEETEBREFTHINSPIRE
DHATHINEVERYHOLTANDHEETHTHETENDRECR
OPPESANDTHEYONGESONNEHATHINTHERAMHI
SHALVECOURSYRONNEANDSMALEFOWELESMAK
ENMELODYETHATSLEPENALTHENYGHTWITHOP
ENYESOPRIKETHHEMNATUREINHIRCORAGEST
HANNELONGENFOLKTOGOONONPILGRIM

Doubly-true (still!) alphametics

By now the hints I was giving earlier are a ghost from the past.

Another hint-free, doubly-true alphametic for you. Don't forget if you're a bit lost, or if you've just opened the book randomly at 41 and thought 'I feel a bit unwell', you can go back to the alphas at 91, 81, 71 (and so on) and try them with the handy hints to get yourself up to speed.

Again, the letters correspond to a unique digit each and the equation holds in word and numerical form. Don't forget the lead digits on each line, the T and F here, cannot equal 0.

We all know that 2 + 14 + 15 = 31. Well, coming in here at number 31 on our countdown, I extend that fact and give you:

```
        TWO
   FOURTEEN
    FIFTEEN
+    TWENTY
-----------
   FIFTYONE
```

Still no hints, sheesh!

But if you're stuck, head on back to the higher numbers (91, 81, 71 and so on) for a refresher.

Get cracking!

30

Feral cats in central Australia have resulted in the extinction of 30 species of mammals

The region has the worst rate of extinction of anywhere in the world. By comparison the entire United States has lost around 90 species of mammal since European settlement (over 400 years ago).

The solution? Birds Australia and the Australian Wildlife Conservancy is building the world's longest cat-proof fence, right alongside the Great Sandy Desert north-west of Alice Springs.

Stage one alone will run 44 kilometres, using 35,000 pickets, 1600 kilometres of plain wire and over 500 kilometres of netting. The 1.8 metre-high fence, when completed, will be up to 10 times as long. If successful it will create a giant sanctuary for native animals free of the scourge of feral cats.

Blankety blanks (#30)

Head on back to number 100 if you need a reminder of how these puzzles work. Rules at the right, pens at the ready ... you know the drill by now ...

6 × (◯ + (◯ − (◯ − ◯) ÷ ◯)÷(◯ − ◯)) = 30

(◯ + (◯ ÷ (◯ − ◯) + ◯) ÷ 6)×(◯ − ◯) = 30

((6 ÷ (◯ − ◯) + ◯) × ◯ − ◯)×(◯ − ◯) = 30

(◯ ÷ (◯ − ◯) + ◯ ÷ (◯ − ◯))×(◯ + ◯) = 30

(◯ − ◯)×(◯ − (◯ + ◯)÷(◯ − ◯))− ◯ = 30

Reach the target number by filling in the blanks in each equation.

A completed equation must incorporate each and every digit from 0 to 9.

You cannot move the operations (+, −, ×, ÷) found between squares.

Order of operations applies!

29

Crowdsourcing

In 1715, the brilliant mathematician and astronomer Edmond Halley (best known for the comet which bears his name) proved that he was also a gun at predicting eclipses.

For the event now known as 'Halley's eclipse' he didn't just predict it would happen, he named the day — 3 May — and the time to within 4 minutes of accuracy.

Further, he asked the people of England to keep an eye out and record what they saw and when they saw it. Using this crowdsourced information, Halley could calculate the eclipse passing over the Earth at 29 miles per minute.

It was just as well the ordinary people did their bit. The two universities of the day were no help at all. According to reports at the time, Dr John Kell at Oxford saw very little 'by reason of clouds' while at Cambridge the Reverend Mr Cotes 'had the misfortune to be oppressed by too much company'!

Dude, you're a professor of astronomy ... there's a solar eclipse which has been predicted by one of the geniuses of the day ... it's happening right outside your window ... and you're feeling oppressed by your company?

Tell them to look out the bloody window at the eclipse!

Sean's Syndesis

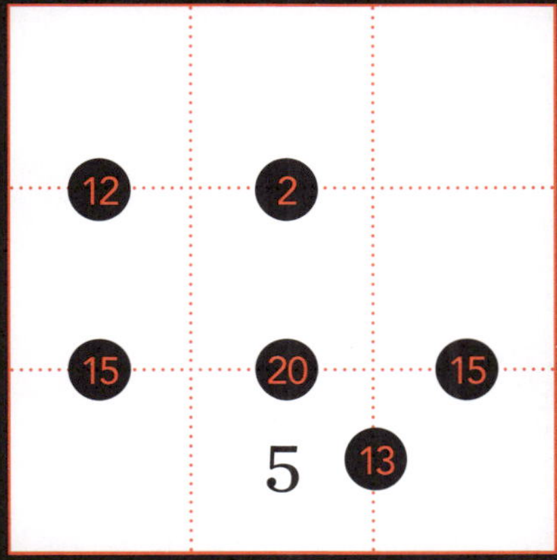

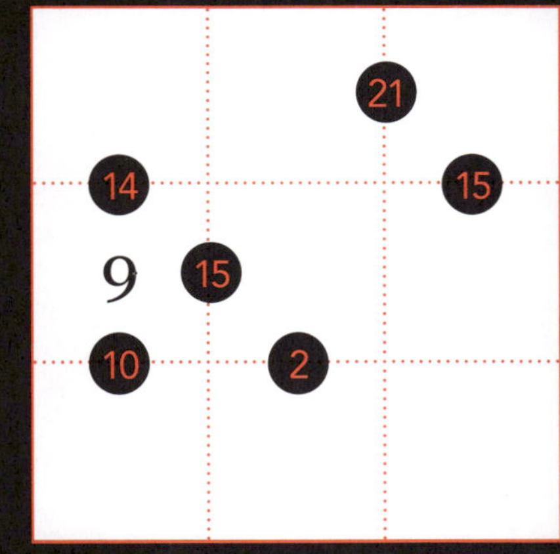

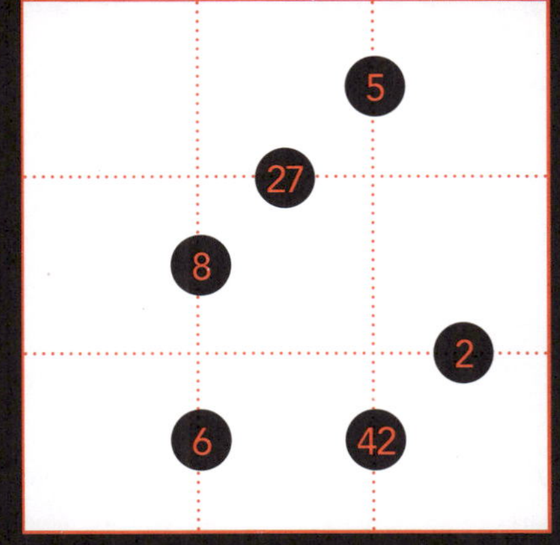

I hope you've enjoyed wrestling with this brand new type of puzzle, the Syndesis, so far in our count-down ... because you're about to attempt your first with no clues at all!

Enter the numbers 1 to 9 into each 3 × 3 grid so that each circled number is the result of adding, subtracting, multiplying, or dividing the two numbers in the cells it touches.

Head on back over to number 99 to read more about these awesome puzzles fresh from the mind of Sean Gardiner!

29

Ffuts gnizama

On Sunday 17 December 2017, the crossword puzzle master (also called a cruciverbalist) for *The New York Times*, the brilliant Will Shortz, broke his own record by creating a monstrous 200-character spiral puzzle for the Christmas puzzle pages of the *NYT*.

Double the length of any previous spiral he had authored, and with 32 inward and 29 outward clues, it's thought to be the biggest ever created.

Okay, this might need some unpacking. First, what is a spiral puzzle?

Unlike a normal crossword where the clues give answers that you read across or down, with a spiral puzzle you start at the outside and spiral in with words following on from each other.

Will Shortz is, fittingly, also the only person known to hold a degree in enigmatology — the study of puzzles — from Indiana University.

It's fair to say he was probably pretty on top of the curriculum, since he wrote it himself.

For example, the 2010 spiral had the following 100 characters as its spiralling inwards answers:

STELLA, BARONESS, WENG, NINEVEH, TOOTER, OFTEN, NO-SPIN, TACTILE, RIFFLES, TISSUES, ROTC, ODESSA, TIME-DELAY, REVE, PACE CAR, GODOT.

'Weng' was a reference to someone's name, 'ROTC' is a military reserves program, 'reve' is French for 'dream' and Godot is the character in a famous play.

So even from these answers it's obviously a pretty tough crossword. But the genius of the spiral is this.

Read all the letters in these answers backwards ... and you get another list of words. Rather than across and down the spiral contains the inward clues and a second set of clues whose answers spiral outwards.

Now, there are some tough ones here, but it works out. Read the letters to the inward answers backwards and you can make the words:

TO DO, GRACE, CAPE VERDE, YALE, DEMITASSE, DOCTOR SEUSS, ITSELF, FIRELIT, CATNIP, SONNET, FORETOOTH, EVENING NEWS, SENORA, BALLETS.

Cape Verde is an Island country off the African coast, 'demitasse' a small coffee and 'senora' is Spanish for 'madam'.

This is *much* harder than creating a traditional crossword and Shortz keeps an amazing list of words in his mind that help to create words in the other direction. For example 'impugning' and 'signing up' in reverse overlap nicely.

Although he readily admits the advent of the internet has made things easier, this is still pure genius, Will Shortz.

28

The mass of Jupiter in kilos is a whopping 28 digits long

Now, 1 million is 7 digits, so what does *28* digits mean?

Say you had 1 kilo. Then imagine, if you really can, 1 million kilos; now how about picturing 1 million blocks of 1 million kilos each; now try to picture 1 million lots of these 1 million blocks of 1 million kilos. Even by now you're only up to 19 digits so you still need 2 billion times as much stuff to hit around the mass of Jupiter. Now try to picture 1,898,130,000,000,000,000,000,000,000 kilos.

I can also tell you that about 1320 Earths would fit inside Jupiter. And that Jupiter is 318 times as massive as Earth (this mismatch between 1320 and 318 is because Jupiter is a gas giant and much less dense than Earth). Jupiter could actually contain all of the other planets in the solar system with room to spare.

One other cool thing about Jupiter is that it is so heavy it doesn't actually orbit the Sun in the way that you might think! While the gravitational attraction of the Sun pulls on Jupiter, Jupiter, with a mass about 1/1000 that of the Sun, is massive enough to exert an observable pull back.

So strictly, Jupiter doesn't orbit the Sun — both the Sun and Jupiter orbit a point just outside the surface of the Sun. Man, that's heavy!

28

Get jiggy with it

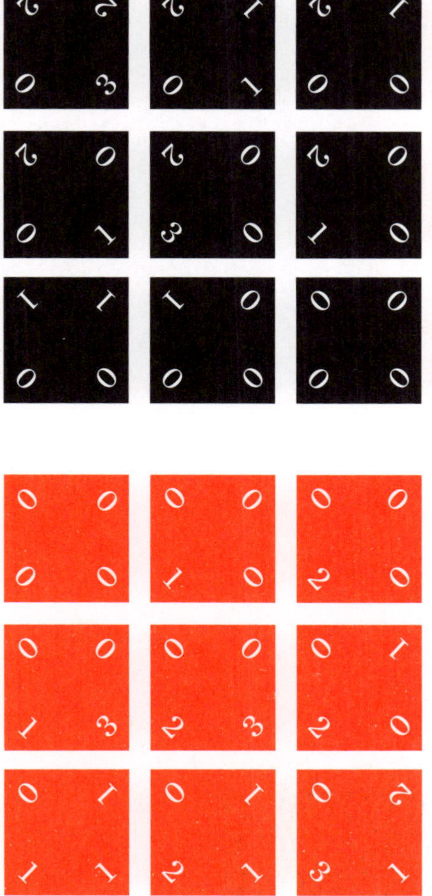

Arrange each set of 9 tiles into a 3 × 3 square so that adjacent tiles show the same numbers along their touching edges. Tiles may need to be rotated, but never reflected. Each set has only one solution. Check number 98 for more information.

For a downloadable cutout, head over to adamspencer.com.au/resources.

27

One of the most controversial pieces of law ever written is just 27 words long

It is the Second Amendment to the Constitution of the United States and reads: 'A well regulated Militia, being necessary to the security of a free State, the right of the people to keep and bear Arms, shall not be infringed.'

Aoccdrnig to rseaerch ...

Form Cmabrigde Uinervtisy, it deosn't mttaer in waht oredr the ltteers of a wrod are. The olny iprmoatnt tihng is taht the frist and lsat ltteers be at the rghit pclae in the wrod. The rset can be a toatl mses and you can sitll raed it wouthit any porbelm at all.

Chances are you've received this email in your inbox over the years. Nifty, eh? But, like many things spammy, it ain't 100% true.

Firstly, there was no Cmabrigde Uinervtisy study. Well, until Matt Davis, a senior research scientist at Cambridge University's Cognition and Brain Sciences Unit, decided to try to track down the origin of this curious tall transposition tale.

Matt discovered that the gist of the email came from a 1999 letter Graham Rawlinson, a child development specialist, had written to *New Scientist*, in response to an article about the effects of reversing short chunks of speech. He noted that the article 'reminds me of my Ph.D. at Nottingham University, which showed that randomising letters in the middle of words had little or no effect on the ability of skilled readers to understand the text'.

Rawlinson and Davis would later go on to put up a website to address the email and explain his research more thoroughly*. 'Clearly, the first and last letters are not the only thing that you use when reading text,' he notes. 'If this were the case, how would you tell the difference between pairs of words like "salt" and "slat"?'

In the text above, 27 of the 55 words — nearly half — are correctly spelled. These serve an important function to help orient us in the sentence.

But dno't jsut tkae my wrod for it — cechk out the wbeiste for mroe ifnomrtaoin. It's a gerat raed.

* You can find the website at mrc-cbu.cam.ac.uk/people/matt.davis/cmabrigde

27

Oh, snap!

As they say in the classics, things just got real.

You and two serious badasses simply can't resolve your differences and agree the only way to settle this is a gunfight. You grab your weapon of choice and head out to the town square and form a triangle, each standing 20 metres away from the pole in the middle of the square upon which the notice of this truel (a 3-way duel) has been posted.

You're a bit nervous, because this really isn't your sort of thing. These two are major players and you're just a humble maths nerd. But at least you've got data on your side. A quick browse of the internet on your way to the square reveals exactly how good a shot these guys are.

Bad guy A (we can't use his name in this book because he has people everywhere and I don't want it to get back to him) is a *seriously* good shot. He hits his target 100% of the time.

Bad girl B (we *definitely* can't use her name – you think A carries a grudge? B can be brutal!) is also a very handy shot and hits her target 80% of the time.

You've handled some heat in your time too, but you know you only hit a target about 27% of the time.

While these are two seriously bad dudes you've wound up having an issue with, at least they have some dignity and play by the rules.

As the worst shot in the group, you get to go first, followed by B (if she's still alive to take a shot) then A (if he's still standing). You'll keep plugging away in that order until one survivor is left.

The percentages and potential strategies are swirling around in your head as the truelmaster (I know, pretty full-on job) tells you that you have 10 seconds to take your shot or you will forfeit your turn.

Clock's ticking ... what do you do?

27

New high score!

Welcome back to the game of High Scoring Equation where you fill in the blanks to win ~~real cash prizes~~ fame and glory! Have fun!

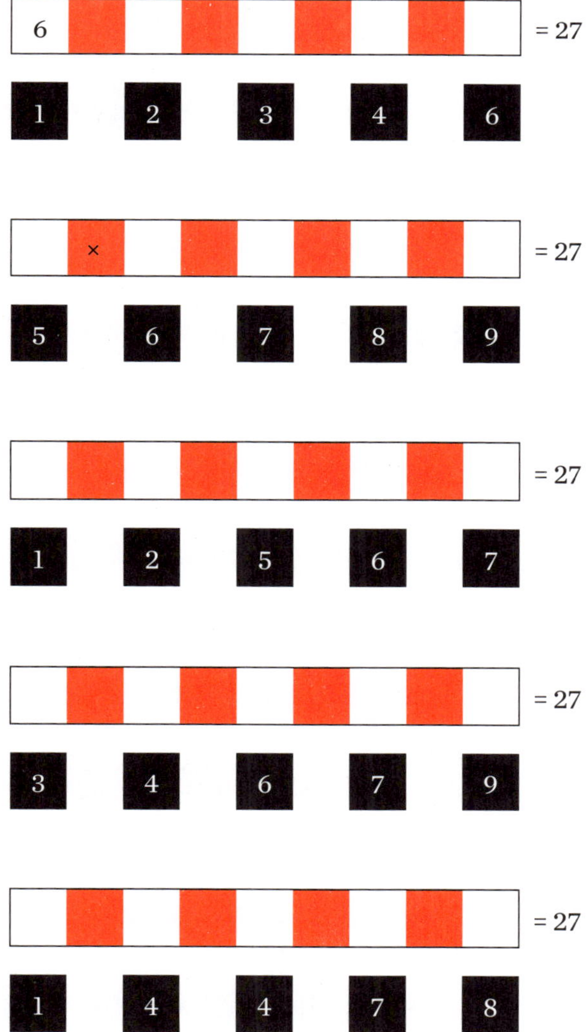

Reach the goal number by creating an equation using the provided numbers and your own choice of operators.

- The numbers should be placed in the white squares.
- Operations (+ – × ÷) are placed in orange squares.
- Order of operations matters, and you can't use brackets!
- Read the numbers off in order for your final score.
- Your goal is to find the equation that gives the highest score.

Head back over to number 97 if you need a refresher on these, otherwise ... get cracking!

In plain Randlish

We know that English has 26 letters in its alphabet. Well, here's a question for you.

Imagine a variation of English called Randlish.

Randlish allows any random jumble of letters to be a word. So, for instance, the word 'CAT' is a word, but so is 'DDFFX' and 'TRETRESEWEWES'.

But Randlish also has a strict letter limit on how long words can be.

How long do you need to allow words to be for there to be more words in Randlish than there are people on the planet?

Go deep

Okay, thrillseekers, strap yourself in for this ride.

This gets a little deep, but I strongly believe it is one of the most beautiful results in all of mathematics that a non-professional mathematician can understand. It takes a bit of work, but we will go in baby steps.

We've spoken a lot about prime numbers – while the number 14 can be broken down to 2×7 which is called the prime factorisation of 14, prime numbers like 7 can be written as 1×7 but not factorised into strictly smaller parts.

Consider the prime numbers 3, 5 and 7. This can be thought of as a 'chain' of 3 prime numbers, starting at 3 and with a difference between them of 2. In fancy-pants maths talk, such a list of numbers with a common difference between them is known as an arithmetic progression and mathematicians write AP-3 to refer to an 'arithmetic progression of primes' of length 3.

So 3, 7, 11 is an AP-3 with difference 4 and 3, 11, 19 is an AP-3 with difference 8.

Can we find chains longer than 3 primes? As former US President Barack Obama said in a very different context ... YES WE CAN!

Turn your eyes away now, grab a pen and paper and find a chain, beginning with the prime 5, that is longer than 3 primes.

...

...

It's very hard to distract you on a printed page, I know. Did you find the AP-4 5, 9, 13, 17? Or perhaps the AP-5 which runs 5, 11, 17, 23, 29. Both of these are arithmetic progressions involving more than 3 primes.

So how long can a chain of primes get? What's the greatest *k* for which we've found an AP-*k*?

Okay, take a breath.

On 18 January 2007 (I know ... *ages* ago) Jarosław Wróblewski discovered a chain of 24 primes. It involves a special number, 223,092,870. This number is special because it is the product of all the primes up to the number 23. As you may recall from number 59, we write this as 23# and refer to it as 'twenty-three primorial'. *

Good ol' Jarosław showed that starting with:

468,395,662,504,823,
and adding 205,619 × 223,092,870 each time,

... you get an AP-24. To find this he used a total of 75 computers working together. Played, JW.

* My hardcore readers ... and I know you're out there ... can prove this to themselves ... *super*-hardcore by pen and paper ... not forgetting that 2 is of course the first prime number.

In 2008, JW and Raanan Chermoni found the first AP-25, starting at:

> 6,171,054,912,832,631 and going up by
> 366,384 × 223,092,870 each time.

I know! My head is spinning too. But we haven't even begun here, my friend. Take another deep breath ... okay ... back in.

The longest chain we have yet found is 26 primes long. It was discovered by Benoãt Perichon on 12 April 2010 ... using ... a PlayStation 3 (I know ... how awesome!) that had been tricked up with JW's prime number-hunting software thanks to a guy called Bryan Little, in a distributed computing project with thousands of people around the world hunting these monsters.

The AP-26 starts at 43,142,746,595,714,191 and goes up by 23,681,770 x 223,092,870 each time.

I can hear you screaming, 'Adam, please, enough!' But it gets so much better!

In 2004, the amazing Aussie mathematician Terence Tao (whom I have gushed about like a fanboy in pretty much every book I've written) and his buddy Ben Green proved that for any whole number *k*, there has to be an AP-*k* out there somewhere in the primes.[1]

Just think about that for a moment.

The longest chain we've found is an AP-26, and that involves seriously big prime numbers.

But the Green–Tao Theorem tells us there must be an AP-27 out there somewhere. It doesn't tell us how to find it, or where to start looking, but we know it must exist.

And there must be an AP-28, an AP-50, an AP-5000 ... yes even an AP-1,000,000,000 must lurk out there somewhere in the endless infinity of prime numbers. For any whole number *k*, no matter how many digits you want to write for it, an AP-*k* must exist.[2]

[1] And if you want to go one step further ... the Green-Tao theorem inherently shows there are infinitely many AP-*k* for any given *k*, so there are infinitely many AP-27 sequences out there, and we can't even find *one*! (For example, there are two AP-27 sequences in each AP-28 sequence, and there are 1,000,000,000 – 26 = 999,999,974 AP-27 sequences in each AP-1,000,000,000 sequence and so on.) I know!!!

[2] I can barely explain how awesome I think that is.

More hexes!

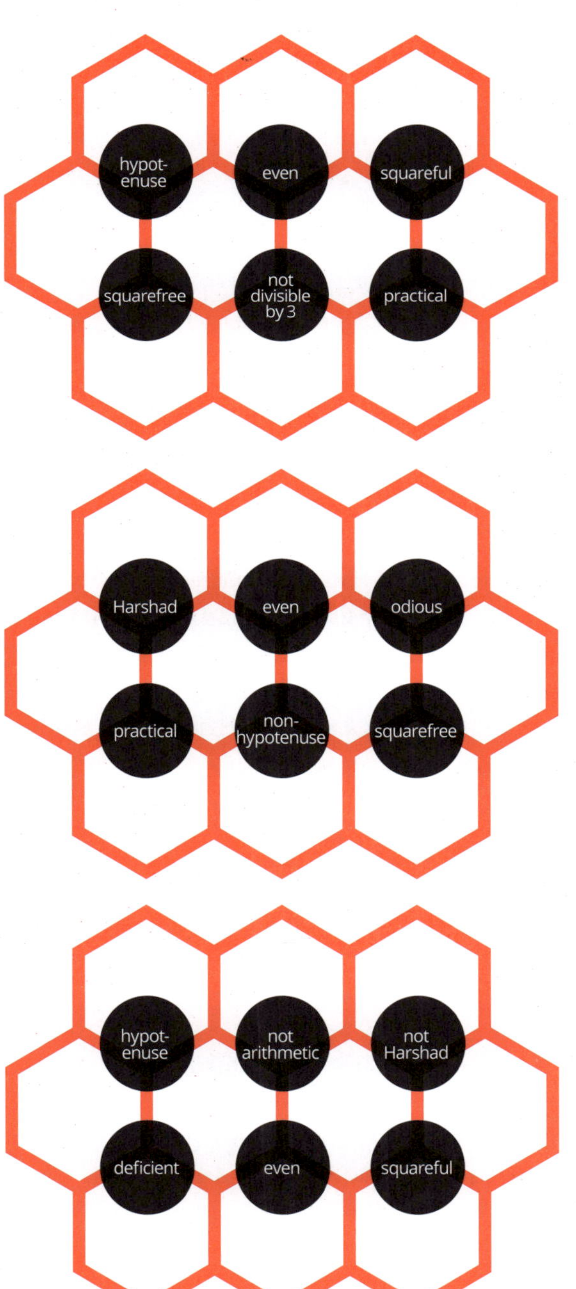

Place the numbers 21 to 30 into each hexagonal grid to satisfy all the rules.

Each cell is touching one or more categories in circles. The number in each cell must belong to all the categories touching it.

Need a hint? If the definitions are a bit tough for you, there's a full list of what these words mean at the end of the book.

26

Some sum, that

One thing we've encountered in this countdown is that you can add an infinite amount of things together and come up with a finite sum. It's a bit of a weird concept when you first hear about it, but if you think of cutting a piece of string in half, then one piece in half again, then one of the smaller pieces in half again and so on ... forever, you can see that $1/2 + 1/4 + 1/8 + 1/16 + 1/32 + ... = 1$.

The infinite amount of terms on the left add together to give us a finite sum.

We write this infinite sum as:

$$\sum_{k=1}^{\infty} \frac{1}{2^k} = 1$$

This terrifying piece of notation ain't really that bad*. It tells us to sum all the terms $1/2^k$ using values $k = 1$, then $k = 2$, then $k = 3$, all the way to infinity. It's a practical way of writing an infinite number of terms in a finite form.

Thanks to the great man himself Leonhard Euler, we have some other cool identities.

$$\sum_{k=1}^{\infty} \frac{k}{2^k} = 2$$

That is, $1/2 + 2/4 + 3/8 + 4/16 + 5/32 + ... = 2$.

$$\sum_{k=1}^{\infty} \frac{k^2}{2^k} = 6 \text{ and } \sum_{k=1}^{\infty} \frac{k^3}{2^k} = 26$$

That is, $1/2 + 4/4 + 9/8 + 16/16 + 25/32 + ... = 6$ and $1/2 + 8/4 + 27/8 + 64/16 + 125/32 + ... = 26$.

Onya, Euler!

* In fact, we already saw this notation at number 96, if you cast your mind back.

26 3 6 4 ... hut!

The numbers 26, 3, 6 and 4 in that order can be used to generate an answer of 26,364 using only the plus, minus, multiplication and division signs, powers and parentheses.

For example, $(26 \times 3 \div 6)^4 = 13^4 = 28{,}561$.

Or $26 + 3^{(6+4)} = 59{,}075$.

Can you make 26, 3, 6, 4 into 26,364?

Check your answer at the back of the book, naturally.

26

And in 26th place ...

The smallest number with exactly 100 divisors is 45,360 = $2^4 \times 3^4 \times 5 \times 7$.

It is the 26th highly composite number.

A number is composite if it can be written as the product of factors in a non-trivial way. So, for example, 10 is composite because 10 = 2 × 5.

The next integer up from 10, namely 11, is not composite because while you can obviously write 11 = 1 × 11, you can't break 11 down into strictly smaller factors.

A number is *highly* composite if it has more factors than any number less than it.

As is often the case with definitions of numbers, things start a bit awkwardly.

The numbers 1 and 2 are both highly composite even though strictly they are not even composite, but they are the first numbers to have 1 and 2 factors respectively. Just go with me on this bit. You should be able to see that 3 with its only factors being 1 and 3 is not highly composite because it only has as many factors as 2 does, whereas 4, with three factors, being 1, 2 and 4, is highly composite.

Question time!

Try to determine all the highly composite numbers up to 100.

Oh, and just in case you were wondering, the 25th highly composite number is 27,720.

25

Marathon efforts

The current world record for running a marathon is 2 hours, 2 minutes, 37 seconds.

That's pretty impressive for covering 42 kilometres and 195 metres (or 26 miles).

Breaking 2 hours for a marathon has been a holy grail for distance runners for decades. Some wonder if it is even humanly possible.

On 6 May 2017, around a Formula 1 race track in Monza, Italy, long distance freak Eliud Kipchoge ran a marathon under conditions designed to make the run as fast as possible.

The race track is flat, other runners took it in turns pacing him alongside and he was wearing the newest-generation race shoes.

He missed out ... by 26 seconds!

2:00:25 is not the new world record because of the artificial race conditions, but still, as we say in Australia, he gave it a red hot go!

25

MERRY XMAS TO ALL

December the 25th is a day when most Aussies share gifts, big lunches and awful dad jokes with those near and dear to them.

But not mathematicians. For real geeks, 25 December is just another day to solve tough number puzzles!

In that spirit, engineer and mathematician Charles Trigg proposed this Christmas curiosity back in 1956 in *American Mathematical Monthly*.

MERRY XMAS TO ALL

'If each letter is a unique representation of a digit, and each word is a square integer, what are these four numbers?'

Try to solve this, noting that there are 10 different letters used here, so the 4 squares cover all the digits 0 to 9.

You don't have to wait till Christmas to check your answer ... you're not too far off the back of the book now!

On the tiles

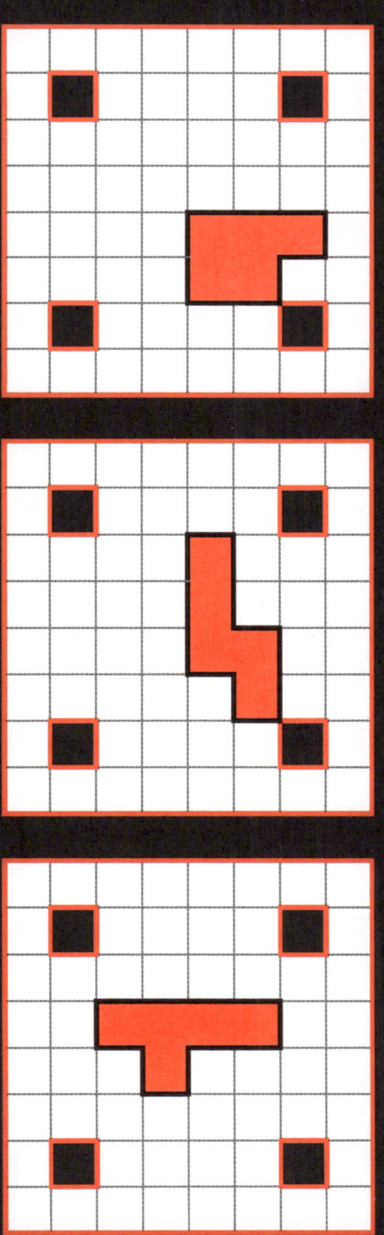

In each of these grids, finish tiling it so that it contains one of each pentomino type.

The 12 different pentominoes are shown here.

You may need to rotate and/or even flip some of them to get them to fit.

Head to adamspencer.com.au/resources if you'd like to download a copy of the puzzle to cut up!

24

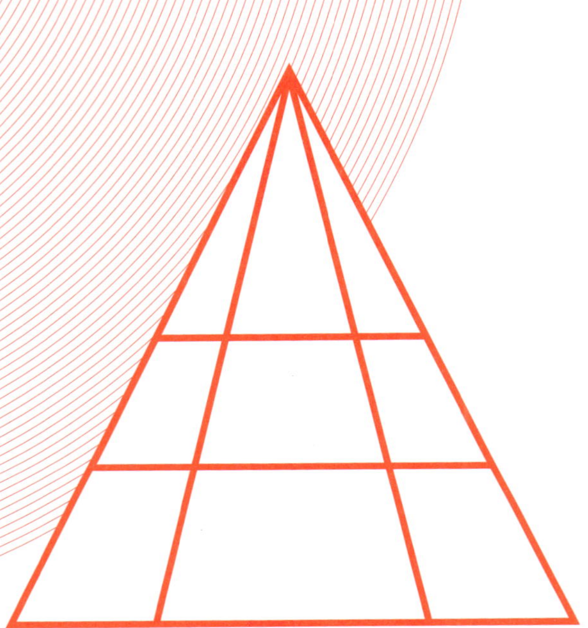

Twenty-four ... more or less?

If you've followed me on social media you'll know the only thing that gets me more excited than an obscure post about number theory, or a video of Buddy Franklin banging home a goal from 65 metres, is an online puzzle that has people stumped!

The question originally asked, 'Four, seven or twenty-four – how many triangles can you count?'

I've posted this puzzle under the number 24, but is that correct?

Anyway, how hard can it be to count up some triangles?

Off you go then. Where in the book should this question actually appear?

Hint — try to think of a logical way to move through the diagram so that you don't miss, or double-count, any triangles.

24

On Christmas Day in 1859 ...

a Geelong farmer named Thomas Austin, also a member of the Acclimatisation Society of Victoria, who had proudly introduced many animals into Australia, including blackbirds and thrushes, decided it would be a good idea to release 24 breeding rabbits onto his property. Some reports suggest he did this to create potential game for shooting parties.

Anyone who has seen the sheer number of rabbits in Australia would suggest, old Tom, that either your mates didn't shoot fast enough or you should have thought things through a bit more!

24

Twenty-four ... hours in a day*

Using the four numbers 3, 3, 8 and 8, and the usual four arithmetic operations (addition, subtraction, multiplication and division), can you make the number 24?

'Ha! Adam, you goose,' you reply, 'that's easy. 3 × 8 = 24. I didn't even need one of your 3s and one of the 8s.'

Your smile holds for about as long as it takes me to say, 'No, you have to use *all four* numbers – both 3s and both 8s.

'No tricks are needed. You can't use 8 and a 3 to make 83, or raise numbers to powers or take square roots or factorials or anything else of that kind.

'Just plus, minus, multiplication and division. Back at ya, playa!'

You can use fractions on the way, of course, because they are the same as division.

But the final answer must be exactly 24. For example: 3 × 8 + (3/8) makes 24 and three-eighths, which is quite close, but no banana.

Trust me ... this one is really tough.

Good luck!

* You might need them for this bad boy!

24

Gridlocks and the three puzzles ...

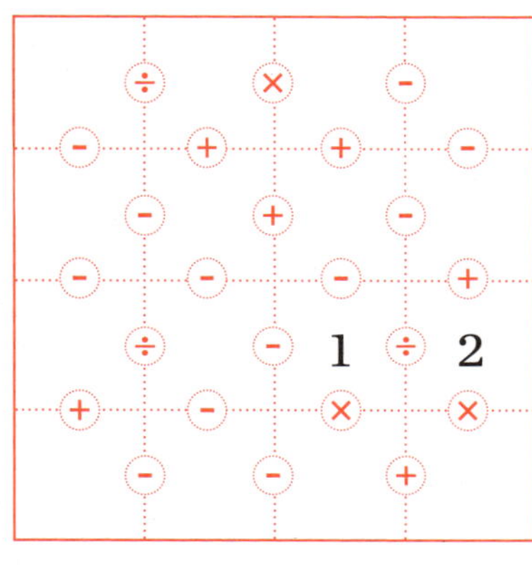

= 2

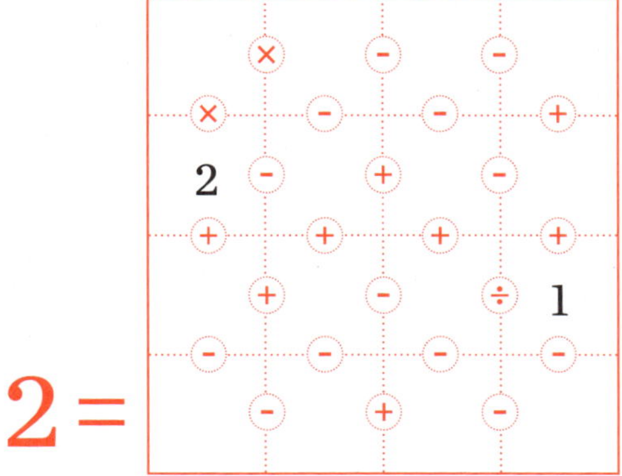

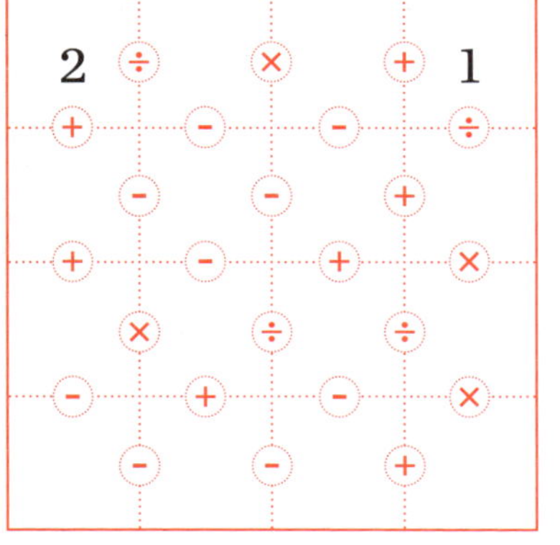

= 2

The rules for these are easy. The answers, well, they're a little tougher!

For each 4 × 4 grid, enter the numbers 1 to 16 so that each row and column equals the target number. As always, order of operations matters.

Can you break the gridlock?

If you don't like defacing this book, go to adamspencer.com/resources for a template to fill in.

23

When grading diamonds by colour we use a 23-step scale

Developed by the GIA (Gemological Institute of America), the scale starts at Z for light yellow or brown diamonds and works up to the most colourless, rare and valuable rocks which score a D.

Why top out at D? There had been many systems before this one, with different scales and amounts of gradings. Many of these schemes used As, Bs and Cs. The new scale was brought in starting at D to avoid any confusion with the old systems.

And yes, darling, I get the hint!

Drill bits (day 8) ...

As we move to the drill run by #23 you can sense the pomp and swagger of 'his Airness' Michael Jordan, who won an amazing 6 NBA Championships with the Chicago Bulls. He is somewhat taken aback when I ask him to resume his place on the bench and welcome the Matildas goal-scoring machine (and Australian soccer superstar) #23 Michelle Heyman to run the session.

In which order did the players stand around the circle?

- #23 stood at the front of the circle.
- #73 stood closer to #93 than to #63.
- There was 1 person between #83 and #3.
- #93 stood closer to #13 than to #73.
- #73, #83, and #3 stood together in some order.
- From #33's perspective, #83 was on the left side of the circle.
- There was 1 person between #63 and #43.
- #23, #53, and #93 stood together in some order.
- From #83's perspective, #23 was on the right side of the circle.

Ten players stood in a circle for their daily drill.

They are numbered 3, 13, 23, 33, 43, 53, 63, 73, 83 and 93.

Each day a different player is the drill leader, which by delightful coincidence, corresponds to the chapter number and 'front position'.

Your task, coach, is to work out the order they stood in the circle each day given the list of clues provided by these fun-loving sporty puzzlers.

Game on!

3

Smashed Avogadro

Twenty-three features in the famous number 6.02×10^{23}.

You're probably thinking, 'Adam, when you say famous ... I've never seen this number before in my life!'

It's a famous number from chemistry. There are about 6.02×10^{23} or 602,214,085,700,000,000,000,000 carbon atoms in 12 grams of carbon-12 (that's about 12 lead pencils). And this number, called Avogadro's number, is used when determining the weight of any substance in chemistry.

In case you're wondering ... and even if you aren't ... one Avogadro's number of people would fill 15 hollowed-out Earths.

Tour de 23

Over 23 days each July the world's best cyclists compete in one of the classic sporting events, the Tour de France.

The 2018 Tour covered over 3500 kilometres over some incredibly hilly terrain. As a result, compared to the 2000 calories an average adult burns per day, a Tour cyclist can melt through 6000 calories.

The *Inside Science* website, run by the American Institute of Physics, calculated this energy demand and worked out if you only ate jelly doughnuts, you'd need to chow through 32 a day to consume 6000 calories. Or if you didn't like jelly doughnuts, perhaps try:

- 9.3 Burger King Whoppers ... or how about
- 24.3 Snickers bars ... or perhaps
- 57.8 bananas ... or maybe you're a
- 242.8 carrots type of person?

The mischievous @qikipedia elves even calculated that to cover the energy expended in doing a Tour de France an average cyclist would have to consume 220 bottles of red wine!

In *Men's Journal,* nutritionist Judith Haudum recommended the following breakfast for an elite cyclist:

- 1 bowl of porridge (150 calories per cup, cooked) with banana (105 calories per banana), and some nuts (529 calories per cup of almonds)
- 1 big plate of pasta (174 calories per cup)
- 1 piece of cake (roughly 225 calories)
- Coffee (1 calorie per cup)
- Fruit juice (122 calories per serving)

And that's just the first of 7 meals ... there's another meal after this before you even start cycling.

23

What's Subba, doc?

Mathematician Subba Rao discovered the following nifty little equation back in 1934:

$$3^6 + 19^6 + 22^6 = 10^6 + 15^6 + 23^6.$$

Note that there are 6 terms here, and the power is also 6.

We don't know if it is the case but the same mathematicians who gave us the gorgeously short maths paper back at 84 teamed up with a guy called John Selfridge in 1967 to 'conjecture' that the total number of terms in this sort of equation has to be at least equal to the power to which all the terms are raised.

If this is true, then you would not be able to find an example where the sum of two 7th powers equalled the sum of three 7th powers. You'd know this couldn't happen because the 2 plus 3 numbers involved here gives 5 total terms, and 5 is less than 7, the power they're being raised to.

But 50 years later this remains a conjecture. We've no idea if it's true or not.

Et tu^{23}?

Nicolaus of Damascus's detailed account of the assassination of Roman Emperor Julius Caesar claims he was stabbed 23 times. Given that there were 60-or-so reported conspirators, it could have been worse, I suppose.

Of the 23 stab wounds, Suetonius (writing 150 years later ... so take it with a grain of salt) reckoned that only one was fatal: the second to his chest which pierced his aorta.

22.92067

Lengthy but awe-sum!

Back at number 80 we looked at the Harmonic Series and the result that 'the sum of the reciprocals of the positive integers that do not contain a 9 sums to less than 80'.

If this phrase makes absolutely no sense to you whatsoever, perhaps you should go back to number 80 and have another look.

We showed the sum has to be less than 80. But what is the sum actually?

In a paper entitled 'Summing Curious, Slowly Convergent, Harmonic Subseries', Thomas Schmelzer and Robert Baillie went to town on a series like this and produced some fascinating results.

As they point out, this stuff is difficult to get your head around for a couple of reasons.

Firstly if:

$$\sum_{n=1}^{\infty} \frac{1}{n} = \frac{1}{1} + \frac{1}{2} + \frac{1}{3} + \frac{1}{4} + \frac{1}{5} + \cdots$$

... goes off to infinity like we know it does, then you'd think that by only removing the terms with 9s in them, for example $1/9$, $1/19$, $1/29$, $1/7986939$ and so on, surely the remaining infinite sum would still go off to infinity.

But as Tom and Bob suggest, best here to quote the mathematician Schumer: 'the problem is that we tend to live among the set of puny integers and generally ignore the vast infinitude of larger ones. How trite and limiting our view!'

The point Schumer is making is that because there aren't that many 9s in the small numbers we usually count up to, we forget that once you get to the 900s, you'll wipe out a whole 100 terms from the series. In fact, once you get to really big numbers, say 50 digits long, the vast bulk of them will contain at least one 9 so they'll all be dropped from the series.

Sure there is an infinite number of terms left in the sum, but we've also taken an infinite number of terms away from the original Harmonic Series and that is what leaves us with a finite sum.

The other limitation on our perception is that these series take a long, long time to get close to their final answer. For the original Harmonic Series, which don't forget goes off to infinity, to reach a sum of even 100 you need to count the first 15,092,688,622,113,788,323,693,563,264,538,101, 449,859,497 terms!

But Tom and Bob have overcome these natural human failings and produced a powerful algorithm for calculating the sum of the Harmonic Series with all the terms removed that contain any string of numbers.

If you remove all the terms that contain the string '42' in them, you get a finite sum of 228.44630 ...

The sum of $1/s$ where s has no even digits is 3.17176 ... If we remove all the terms with odd digits the sum of the remaining terms is smaller still at 1.96260 ...

What about if you remove any number that contains the first 3 digits of π, namely '314'? Obviously a lot more terms remain and the sum is larger: 2299.82978 ...

Perhaps the first 6 digits of π; remove all occurrences of '314159' and the sum is: 2302582.33386 ...

Okay, Adam, you've had your fun. But wait, there's more!

So the final take-away here is that while back at number 80 we had a neat proof that the sum of the Harmonic Series with all 9s removed had to be less that 80, the real sum is

22.92067

much smaller. To 5 decimals, that sum is 22.92067 ...

This table neatly summarises the sums to the first 5 decimal places of the Harmonic Series with all terms containing each individual digit removed:

0	23.10344
1	16.14769
2	19.25735
3	20.56987
4	21.32746
5	21.83460
6	22.20559
7	22.49347
8	22.72636
9	22.92067

Yeah, baby.

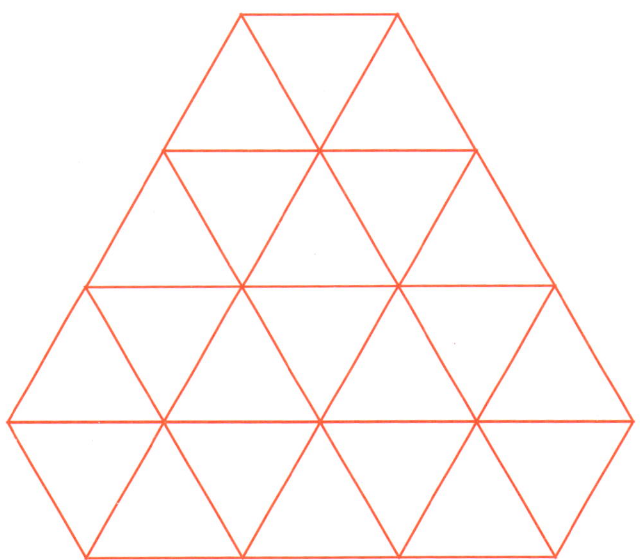

Regal rooms

James Tanton, Assistant Professor of Mathematics at St. Mary's College of Maryland, is an Aussie and one of Twitter's most prolific maths puzzlers.

In 2017, his 'Exploding Dots' project engaged over 1.7 million students around the world as part of the first ever Global Math Week. Great work, James!

JT posed this little number back in 2002.

A king lives in this 22-room palace. Every room is a triangle. Before retiring for the night, he likes to inspect his palace, room by room.

Is there a path that will let him visit each room once and *only* once? He can start anywhere ... as can you!

Off you go!

... and as a hint ... bonus points if you can solve this without even using a pen on the page.

22

Twenty-two is typically the minimum number of episodes per season on US TV

After the first or 'pilot' epside, if the show is picked up, the network (kids, ask your parents ... it's a pre-Stan thing ... will order a 'run' of episodes (usually 6 or 13 to start with), however a typical season is a full 22 episodes.

22

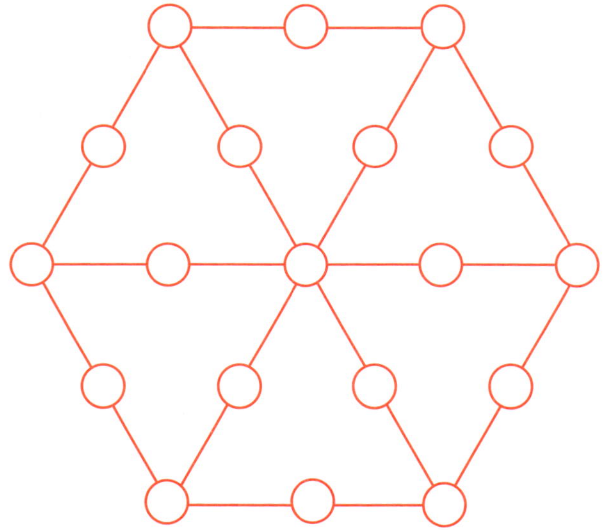

A hex(agon) upon you

While perusing the amazing puzzles at plus.maths.org one day I came across this little beauty which 'Marianne' posted back on 1 September 2016.

There are 19 dots arranged in a hexagon.

Your task is to label the dots with the numbers 1 to 19 so that each set of the 3 dots that lie along a triangle's side add up to 22.

As Marianne herself said ... happy puzzling!

Need to find the area of an old regular hexagon you have lying around the house?

Easy-peasy: $3\sqrt{3}/2 \times (\text{side})^2$

22

An interesting fact(orial)

Twenty-two factorial has 22 digits. Don't believe me?

22! = 1,124,000,727,777,607,680,000

There are the only 3 cases where n! has n digits.

We know this because once we get to 25! = 15,511,210, 043,330,985,984,000,000 which has 26 digits, with each new factorial we are multiplying by more than 10 thus adding at least one more digit to the answer.

This illustrates how quickly factorials increase as we saw with the awesome example of 100! earlier.

What's that? What are the other two cases? I'm very glad you asked:

23! = 25,852,016,738,884,976,640,000 has 23 digits and
24! = 620,448,401,733,239,439,360,000 has 24 digits.

Snakes alive!

→

9	–	6	–	9
+	5	+	3	–
5	×	3	×	3
×	6	×	8	=
5	×	6	=	22

→

3	–	6	×	6
–	5	×	9	+
6	+	6	–	7
–	4	–	6	=
7	×	1	=	22

→

3	×	4	×	3
–	7	×	5	×
8	×	6	–	2
–	3	–	3	=
6	×	4	=	22

In these grids, you can make paths from the top-left corner to the bottom-right by moving between adjacent squares.

You can move up, down, left or right, but you can never visit the same square twice.

And you guessed it, the paths must trace out the correct equation to reach the target number in the bottom right-hand corner.

For each grid, there are 3 different paths that all trace out correct equations.

Head on back to number 92 (if you haven't been there already) if you need a refresher.

21

Doubly-true (still!) alphametics

Only 3 of these amazing little sums to go!

Our doubly-true alphametic for 21 involves both 20 and 1 and is:

```
  EIGHT
  EIGHT
    TWO
    ONE
+   ONE
-------
 TWENTY
```

Again, the letters correspond to a unique digit each and the equation holds in word and numerical form.

Don't forget the lead digits on each line, the E, O and T here cannot equal 0.

Good luck.

Nothing to see here, I'm afraid ...

The human soul weighs 21 grams

So says Duncan MacDougall in his 1907 study on the matter.

He hypothesised that souls have a physical weight, and set about trying to prove it. He chose 6 patients in nursing homes whose death was obviously imminent (4 tuberculosis sufferers, 1 diabetic and 1 ... unspecified dying person) and placed their entire beds on industrial scales sensitive to within 5.6 grams.

One of his subjects lost ¾ of an ounce, or 21.3 grams of weight ... and MacDougall's infamy was as assured as his certainty that he'd measured the weight of a soul.

Obviously his woefully small sample size, unscientific methods and the fact that only a single subject met his hypothesis render his research meaningless.

But good on you for having a crack, MacDougall.

20

Here's some luck for you ...

You might need it.

I first saw this question asked on Twitter by Paul Godding (@7puzzle) back in 2015.

'Use 1, 5, 5 and 10 once each, with + – × ÷ available, to arrive at even numbers 2—20, but are any impossible? 2 is possible: $[(5+5) \div 10] + 1 = 2$'

Of course this got me thinking.

Why not try to get the other even numbers 4, 6, 8 ... up to 20?

And when you're done, why not see what odd numbers you can achieve?

Finally, if I allowed you a square root sign as well, so $\sqrt{5-1} = 4$ came into play, what could you get?

Remember, you must use exactly one 10, two 5s and a 1 in your equation.

Good luck! Check your findings at the back of the book.

Blankety blanks (#20)

Head on back to number 100 if you need a reminder of how these puzzles work. I've only give you one clue here, but I reckon you're probably pretty up on these by now. Rules at the right, pens at the ready ... get solving!

$$((\bigcirc + \bigcirc)\times(\bigcirc + \bigcirc)\div(\bigcirc - \bigcirc) - \bigcirc) \div 4 = 20$$

$$(9 \times (\bigcirc - \bigcirc) - \bigcirc)\div(\bigcirc \times (\bigcirc - \bigcirc) - \bigcirc) = 20$$

$$\bigcirc \times (\bigcirc + \bigcirc)\div(\bigcirc - (\bigcirc - \bigcirc)\times(\bigcirc - \bigcirc)) = 20$$

$$(\bigcirc + \bigcirc)\times(\bigcirc - (\bigcirc + \bigcirc)\div(\bigcirc + \bigcirc) - \bigcirc) = 20$$

$$(\bigcirc \times((\bigcirc + \bigcirc)\div(\bigcirc + \bigcirc) + \bigcirc) - \bigcirc) \times \bigcirc = 20$$

Reach the target number by filling in the blanks in each equation.

A completed equation must incorporate each and every digit from 0 to 9.

You cannot move the operations (+, −, ×, ÷) found between blanks.

Order of operations applies!

20

Rinse and repeat (1,048,576 times)

The number $919{,}444^{2^{20}} + 1$ ... is prime.

Not just any prime. This beast is 6,253,210 digits long, and the 13th largest prime we currently know. So what's the largest prime we currently know? Well, it makes our humble $919{,}444^{2^{20}} + 1$ look like a minnow.

$2^{77{,}232{,}917} - 1$, which was discovered the day after Christmas 2017 (what a lovely gift, geek-Santa!) is a mindboggling 23,249,425 digits long.

If the form of this scares you, here is what it means.

First, 2^{20} is obtained by multiplying twenty 2s together, giving us:

$$2 \times 2 \times 2 \times 2 \times 2 \times 2 \times 2 \times 2 \times 2 \times 2 \times 2 \times 2 \times 2 \times 2 \times 2 \times 2 \times 2 \times 2 \times 2 \times 2 = 1{,}048{,}576$$

So $919{,}444^{2^{20}} + 1 = 919{,}444^{1{,}048{,}576} + 1$, meaning we just write down the number 919,444 a measly 1,048,576 times ... then multiply them all together. Easy!

Simply add 1 and you'll find yourself with the 13th largest prime number.

Ever wondered why we call prime numbers, well, 'prime'?

If you're still wondering, you're obviously not reading this book in order. Fair enough.

Head on back to number 70 for a bit of a history lesson featuring the (extremely) late, great Euclid ...

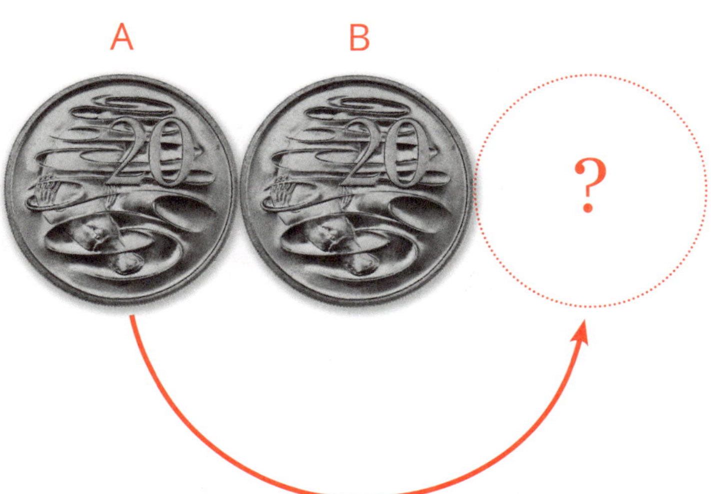

Rock and (20c) Roll!

I place two 20 cent coins next to each other so that they are touching and both with the number 20 written upwards.

If I keep coin B fixed in place and roll coin A around it, going underneath coin B, when it gets to the other side will the 20 be written up or down?

And if I rolled coin A over the top of coin B would the answer be any different?

Have a think, then go get a couple of coins and give it go. Check your findings at the back of the book.

The Aussie 20 cent piece with the iconic platypus design was sculpted by Stuart Devlin and introduced all the way back in 1966.

19

The first McDonald's is often credited as being opened in San Bernardino, CA on 15 May, back in 1940

But strictly speaking, the McDonald brothers Dick and Mac opened their first eatery next to Southern California's Monrovia Airport in 1937. This tiny octagonal building was later moved to 1398 North E Street in San Bernardino, California and reopened in 1940.

The production-line cooking method allowed the brothers to sell burgers for only 15 cents, when most competitors charged more like 30 cents. But if you were feeling flashy, they did offer what they named the 'Tempting Cheeseburger' for a whole ... 19 cents.

Sean's Syndesis

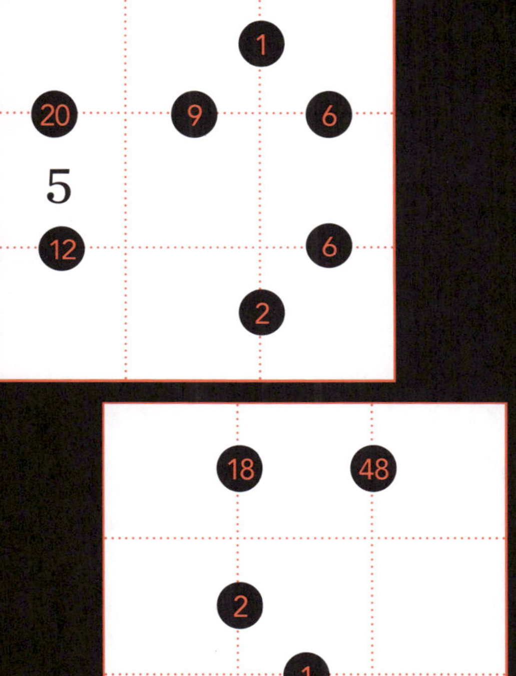

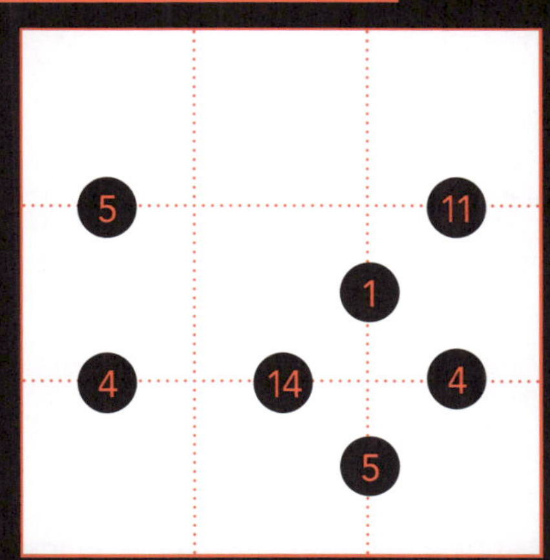

Enter the numbers 1 to 9 into each 3 × 3 grid so that each circled number is the result of adding, subtracting, multiplying, or dividing the two numbers in the cells it touches.

Just 1 clue now for you to kick things off.

Head on back over to number 99 to read more about these awesome puzzles fresh from the mind of Sean Gardiner!

18

Suosikki sanaani suomeksi

If you can speak a bit of Finnish, you don't need me to tell you that the word 'hyppytyyny' means 'bouncy cushion'.

Half of this entire, awesome, very specific word is made up of the letter 'y'!

Further, to those of you who can speak Finnish, and I know that's a very small cross section of my lukijoita,* you'll also be familiar with my favourite Finnish word ... at the moment.

Brace yourself for 18 letters of pure joy:

hyppytyynytyydytys

Gorgeous. Oh, what does it mean?

'Bouncy cushion satisfaction'.

May your day be full of hyppytyynytyydytys.

* Readers.

The jiggles

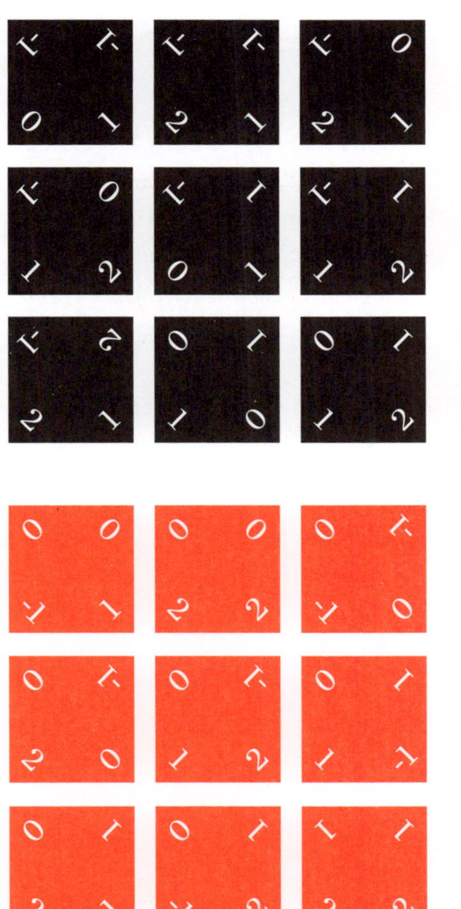

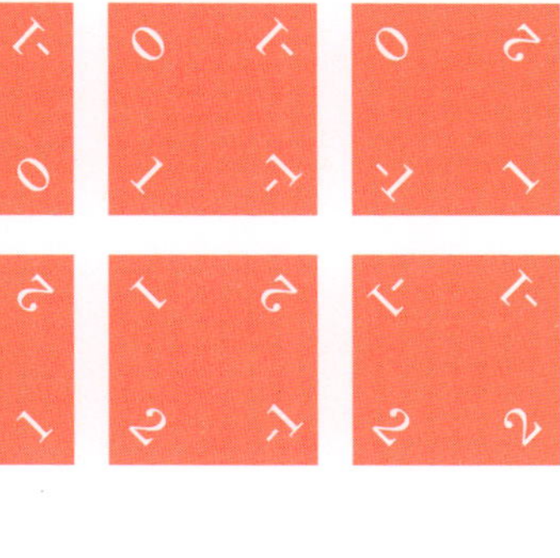

Arrange each set of 9 tiles into a 3 × 3 square so that adjacent tiles show the same numbers along their touching edges.

Tiles may need to be rotated, but never reflected. Each set has only one solution. Oh, and note that we've had to introduce negative numbers at this point for the board totals to equal 18 ... have fun!

For a downloadable cutout, go to adamspencer.com.au/resources.

17

Prime obsession

Continuing my obsession with large prime numbers, the number 17 features in a delicious way.

First, way back in 1951, Aimé Ferrier showed that the number

$$\frac{2^{148}+1}{17}$$

is prime. Written out in its full glory it is 20,988,936,657, 440,586,486,151,264,256,610,222,593,863,921 and at 44 digits long, IMHO a beautiful prime for the following reason: Ferrier cracked this nut using a mechanical calculator! It is the only massive prime discovered this way. The previous largest prime we knew, the 39-digit

$$2^{127} - 1 = 170{,}141{,}183{,}460{,}469{,}231{,}731{,}687{,}303{,}715{,}884{,}105{,}727$$

... was shown to be prime by the brilliant prime-slayer Édouard Lucas and he did it *by hand*.

And only months after Ferrier's effort, the EDSAC computer at Cambridge University took Lucas's prime, which we abbreviate as M_{127} and messed around with it a bit to show that

$$180 \times (M_{127})^2 + 1 \text{ is a 79-digit prime.}$$

From this moment on it would be computers that set all the prime discovery records. But Ferrier's prime remains a lovely snapshot of that brief in-between time when a mechanical device held sway.

New high score!

Welcome back to the game of High Scoring Equation where you fill in the blanks to win ~~real cash prizes~~ fame and glory! Have fun with you last clue!

		3							= 17

3	3	6	7	8

									= 17

2	4	7	8	9

									= 17

3	4	5	7	9

									= 17

5	6	7	8	9

									= 17

4	6	6	7	9

Reach the goal number by creating an equation using the provided numbers and your own choice of operators.

- The numbers should be placed in the white squares.
- Operations (+ – × ÷) are placed in orange squares.
- Order of operations matters, and you can't use brackets!
- Read the numbers off in order for your final score.
- Your goal is to find the equation that gives the highest score.

Head back over to number 97 if you need a refresher on these, otherwise ... get cracking!

The French flag was once entirely white

Yep. For 16 years, in fact, from 1814 until 1830.

16

More hexes!

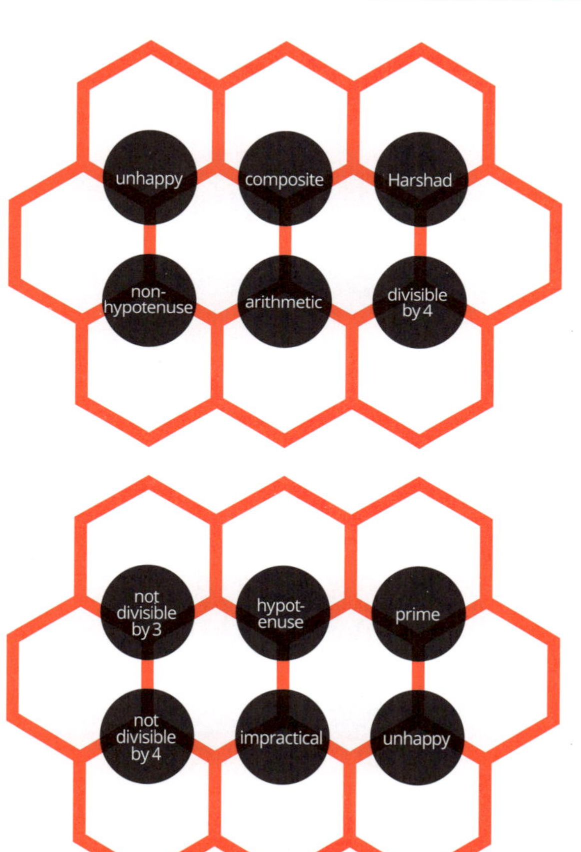

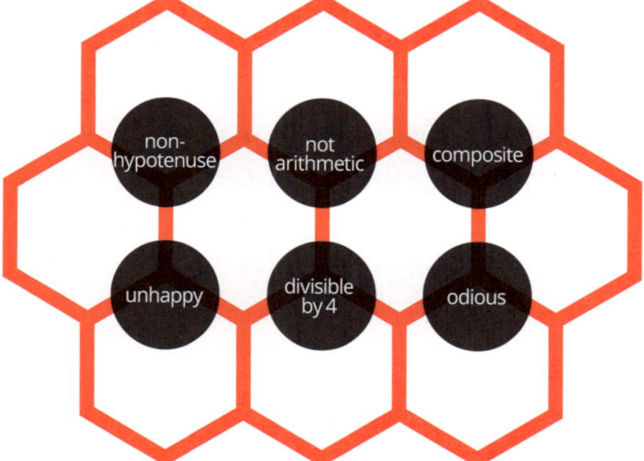

Place the numbers 11 to 20 into each hexagonal grid to satisfy all the rules.

Each cell is touching one or more categories in circles. The number in each cell must belong to all the categories touching it.

Need a hint? If the definitions are a bit tough for you, there's a full list of what these words mean at the end of the book.

15

Un©able

What's the longest word in the English language?

As the good folks over at the Oxford English Dictionary say, for many of the contenders 'you're unlikely to come across in genuine use: they're generally just provided as answers to questions about the longest words in the English language'. They note that the current title holder – pneumonoultramicroscopicsilicovolcanoconiosis – with its 45 letters is a 'a *supposed* lung disease'. Italics added. Burn, OED.

Less contentious is the question of the longest English word which doesn't repeat a letter.

The title for this gem is tied: uncopyrightable is one word. The other? Dermatoglyphics, meaning 'the study of skin markings'.

The longest word to be found in Britain is the Welsh place name Llanfairpwllgwyngyllgogerychwyrndrobwllllantysiliogogogoch.

Look, Welsh is a tough language to pick up by most standards, but this takes the cake.

If you decide to visit, you may be pleased to know that the name is often abbreviated to 'Llanfair PG'.

On the tiles

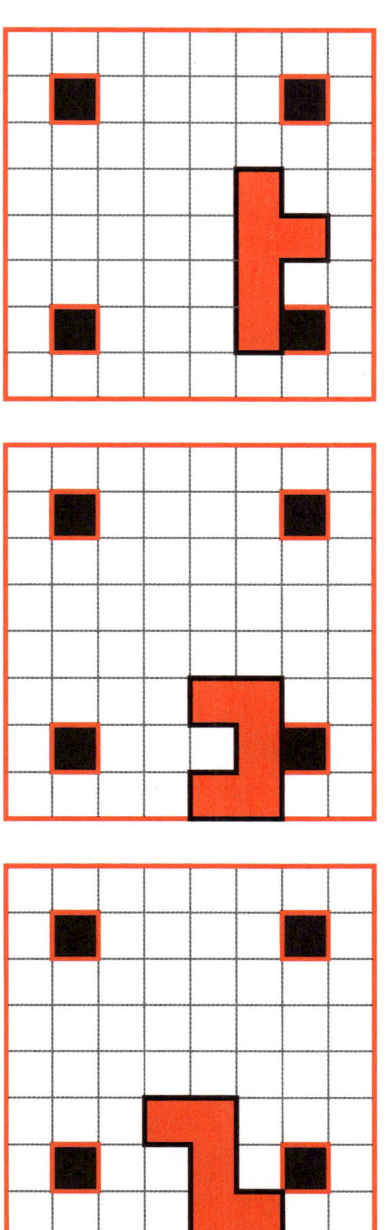

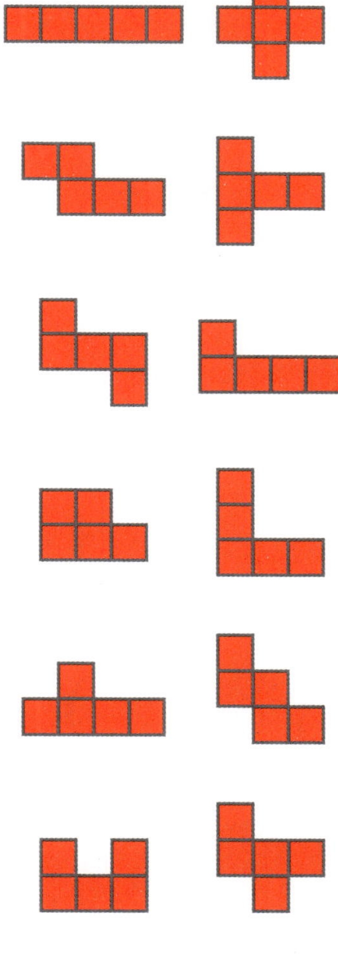

In each of these grids, finish tiling it so that it contains one of each pentomino type.

The 12 different pentominoes are shown here.

You may need to rotate and/or even flip some of them to get them to fit.

Head to adamspencer.com.au/resources if you'd like to download a copy of the puzzle to cut up!

14

Don't go changin'

You've heard the phrase 'like watching paint dry' or 'like watching grass grow'?

How about watching 12 extremely accurate clocks tick over ... for around 14 years*!

* ... or roughly 450 million seconds, if you prefer ...

That's exactly what Bijunath Patla from the National Institute of Standards and Technology and his team did.

He monitored these atomic clocks, powered by two different types of atoms, kept in vacuum-sealed, temperature and humidity controlled conditions, as the Earth revolved around the Sun, in order to prove that the laws of physics are the same no matter where or when you are in the universe.

As tedious as it may seem it's actually really important in understanding the nature of the universe we live in today to know if the rules have changed, even slightly.

Well, good old BP and the team say that this incredible exercise in intellectual tedium gives weight to the claim that yes, the laws of physics are unchanging.

Beat the gridlock!

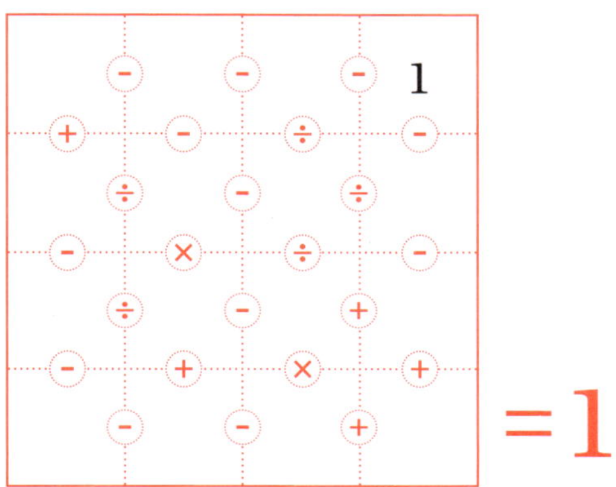

=1

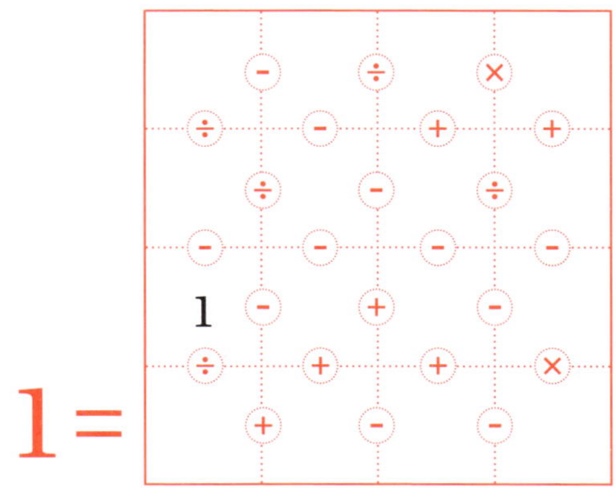

1=

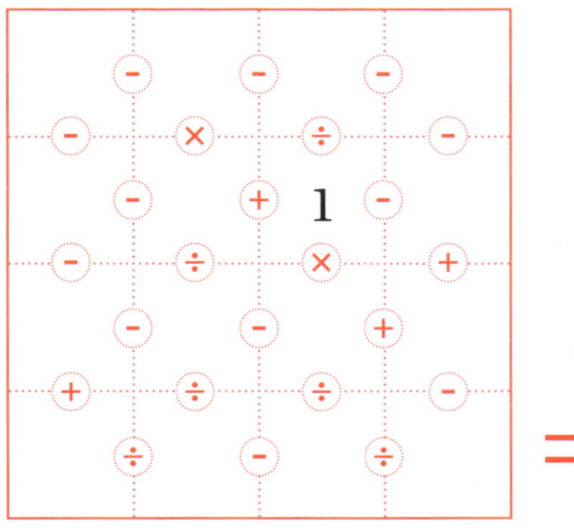

=1

The rules for these are easy. The answers, well, they're a little tougher!

For each 4 × 4 grid, enter the numbers 1 to 16 so that each row and column equals the target number. As always, order of operations matters.

Can you break the gridlock?

If you don't like defacing this book, go to adamspencer.com/resources for a template to fill in.

13

High times

Let's revisit some high school multiplication. You know, the old 'put down the 4, carry the 2' type stuff.

I want you to calculate:

$13 \times 53^2 \times 3853 \times 96179$.

Go on, trust me ... it's worth it. The answer isn't at the back of the book, it crops up again in just a few, well, paragraphs. Go on!

Okay, who did their homework all the way back in the last paragraph?

If you crunched that out correctly you've probably already realised the awesomeness that I first discovered on the murderousmaths.co.uk website administered by my old mate Kjartan Poskitt. Introducing the fabulous ...

13532385396179

There are many numbers for which this 'concatenation' property holds.

What's impressive about this one and made it worthy of Conway's $1000 prize is that the number is the concatenation of its prime factorisation (in order) — that is: 13, 53, 3853, and 96179 are all prime. Awesome.

Props to James Davis who recently won $1000 by discovering that:

$13532385396179 = 13 \times 53^2 \times 3853 \times 96179$.

Look at the digits ... 13 53 2 and so on. The same sequence of numbers occurs in both the answer and the sums!

That is AWESOME!

Sure you could argue in the greater scheme of things it is also pointless, but that is just being a spoilsport as far as I'm concerned.

Drill bits (day 9)

This second-last skills session will be taken by #13, the great Erin Phillips from the Adelaide AFLW football club.

Erin is the former Australian basketball legend who returned to the game of AFL in Australia's first-ever national womens' league and promptly won the awards for best player of the season, best player in the grand final and goal of the year! #Legend

In which order did the players stand around the circle?

- #13 stood at the front of the circle.
- #23, #73, and #3 stood together in some order.
- #63 and #33 stood the same distance from #73.
- #43, #93, and #3 stood together in some order.
- There were 2 people between #3 and #13.
- #43, #53, and #63 stood together in some order.
- From #73's perspective, #93 was on the left side of the circle.

Ten players stood in a circle for their daily drill.

They are numbered 3, 13, 23, 33, 43, 53, 63, 73, 83 and 93.

Each day a different player is the drill leader, which by delightful coincidence, corresponds to the chapter number and 'front position'.

Your task, coach, is to work out the order they stood in the circle each day given the list of clues provided by these fun-loving sporty puzzlers.

Game on!

12

The hands of a clock are on top of each other at 12 midnight

Now, you may have already figured that one out.

Okay then, how many times during the day will they sit on each other and at what time (to the nearest second) will this *first* happen?

Snakes alive!

→

1	–	4	–	7
–	1	+	5	×
2	+	8	–	3
–	3	–	8	=
1	–	1	=	12

→

3	–	9	+	2
–	2	×	5	–
2	–	7	–	3
×	4	–	1	=
3	+	7	=	12

→

6	+	2	×	9
×	7	×	3	–
9	+	9	+	3
×	4	–	1	=
4	+	2	=	12

In these grids, you can make paths from the top-left corner to the bottom-right by moving between adjacent squares.

You can move up, down, left or right, but you can never visit the same square twice.

And you guessed it, the paths must trace out the correct equation to reach the target number in the bottom right-hand corner.

For each grid, there are 3 different paths that all trace out correct equations.

Head on back to number 92 (if you haven't been there already) if you need a refresher.

12

Batten the hatches!

Ahoy, me hearties!

The sea she can be a cruel mistress. Especially when the Beaufort scale of winds gets up towards its maximum of 12. Here 'tis, in all its glory ...

0 | Calm

1 | Light air at 1–5 km/h. Expect 0–0.2 metre waves. Ripples with appearance of scales are formed, without foam crests.

2 | Light breeze at 6–11 km/h. Expect 0.2–0.5 metre small wavelets – still short but more pronounced. Crests have a glassy appearance but do not break. Wind felt on face; leaves rustle; wind vane moved by wind.

3 | Gentle breeze at 12–19 km/h. Expect 0.5–1 metre large wavelets. Crests begin to break; foam of glassy appearance; perhaps scattered white horses. Leaves and small twigs in constant motion; light flags extended.

4 | Moderate breeze at 20–28 km/h. Expect 1–2 metre small waves becoming longer; fairly frequent white horses. Raises dust and loose paper; small branches moved.

5 | Fresh breeze at 29–38 km/h. Expect 2–3 metre moderate waves taking a more pronounced long form; many white horses are formed; chance of some spray. Small trees in leaf begin to sway; crested wavelets form on inland waters.

6 | Strong breeze at 39–49 km/h. Expect 3–4 metre large waves begin to form; the white foam crests are more extensive everywhere; probably some spray. Large branches in motion; whistling heard in telegraph wires; umbrellas used with difficulty.

7 | High wind, moderate gale, near gale at 50–61 km/h. Expect 4–5.5 metre seas heaping up, and white foam from breaking waves beginning to be blown in streaks along the direction of the wind. Spindrift* begins to be seen. Whole trees in motion; inconvenience felt when walking against the wind.

8 | Gale, fresh gale at 62–74 km/h. Expect 5.5–7.5 metre, moderately high waves of greater length. The edges of crests break into spindrift; foam is blown in well-marked streaks along the direction of the wind. Twigs break off trees; generally impedes progress.

9 | Strong/severe gale at 75–88 km/h. Expect 7–10 metre high waves with dense streaks of foam along the direction of the wind; sea begins to roll; spray affects visibility. Slight structural damage (chimney pots and slates removed).

10 | Storm, whole gale at 89–102 km/h. Expect 9–12.5 metre, very high waves with long overhanging crests with resulting foam in great patches blown in dense white streaks along the direction of the wind. On the whole, the surface of the sea takes on a white appearance; rolling of the sea becomes heavy; visibility affected. Seldom experienced inland. Trees uprooted; considerable structural damage.

11 | Violent storm at 103–117 km/h. Expect 11.5–16 metre, exceptionally high waves. Small- and medium-sized ships might be lost to view behind the waves for a long time. Sea is covered with long white patches of foam; everywhere the edges of the wave crests are blown into foam; visibility affected. Very rarely experienced; accompanied by widespread damage.

12 | Hurricane force ≥ 118 km/h. Expect ≥ 14 metre waves. The air is filled with foam and spray; sea is completely white with driving spray; visibility very seriously affected. Devastation.

* What's 'spindrift', you ask? Why, it's spray blown from cresting waves during ... a gale.

11

The US Susan B. Anthony dollar is a rare example of an undecagon coin

AKA a hendecagon, or an 'awesome 11-sided regular polygon'!

S.B.A. was an American women's rights activist who formed various equal rights associations, published a women's rights magazine and generally caused uppity 'female-type' trouble for sexist white blowhards in the late 19th century.

In 1872, she was arrested for voting in her hometown of Rochester, New York (and if you close your eyes and listen closely you can still hear dinosaurs saying 'rightly so'). Many say this got the ball rolling on what was signed into law as the 19th Amendment to the US Constitution prohibiting the states and the federal government from denying the right to vote to citizens on the basis of sex ... on 18 August 1920.

Yep ... just *48 years* later!

Doubly-true (still!) alphametics

For our second-last alphametic we go back to the future to the example I showed you way back when.

The doubly-true alphametic with the lowest possible answer that has only one solution.

Back at 91, I set you the challenge of answering this but gave you the value of THREE to get you started. That's all lost back in the mists of time now, so I challenge you to answer this same alphametic ... without hints!

```
  THREE
  THREE
    TWO
    TWO
+   ONE
-------
 ELEVEN
```

Good luck!

Nope, still no hints!

10

The word 'decimate' comes from the number 10

Yup. Though it's often misused as a synonym for 'destroy', the correct meaning of the verb 'to decimate' is 'to reduce by a 10th'.

Its usage can be traced all the way back to ancient Rome, where cowardice or mutiny could be punished by the execution of 1 in 10 soldiers in a cohort.

Certainly not great for 1 in 10 of the crew involved, but a fair bit short of complete destruction.

#ancientrome #bloodthirsty #verybloodthirsty #whoknew?

10

Sweet geeky dreams ...

I like to think I've had some pretty messed up dreams in my life.

Once, as a child, I can remember falling in slow motion down a giant hole in the ground with massive sets of hands that closed together as if to save me, only to open at the last second as an angelic voice sang, 'Sometimes you catch them, sometimes you don't'.

Anyway, enough of my therapy session. I'm sure you've also had your share of weird dreams.

But how do you stack up with the great Indian mathematician Ramanujan? The subject of the movie *The Man Who Knew Infinity,* whose life was cut tragically short in his early thirties, said that it was in his dreams that he came up with mathematical formulae like this:

$$10 - \pi^2 = \sum_{k=1}^{\infty} \frac{1}{k^3 (k+1)^3} = \frac{1}{1^3 \times 2^3} + \frac{1}{2^3 \times 3^3} + \frac{1}{3^3 \times 4^3} + \frac{1}{4^3 \times 5^3} + \ldots$$

I don't know what you need to eat before bedtime to dream of this sort of thing, but get me a bowl of the stuff!

You might want to bang around with a calculator and find out this: in the equation above, how quickly does the expression on the right converge to match the value on the left to 4 decimal places?

10

Blankety blanks (#10)

Head on back to number 100 if you need a reminder of how these puzzles work. Not a single solitary clue now, but I have faith in you! Rules on the left (just to mix things up), pens at the ready ... hit it!

$$(\bigcirc - \bigcirc)\times(\bigcirc - (\bigcirc + \bigcirc - \bigcirc)\times(\bigcirc - \bigcirc)) = 10$$

$$((\bigcirc \times (\bigcirc + \bigcirc) - \bigcirc)\div(\bigcirc - \bigcirc) - \bigcirc) \div \bigcirc = 10$$

$$(\bigcirc \times (\bigcirc + \bigcirc) - \bigcirc \div (\bigcirc - \bigcirc))\div(\bigcirc - \bigcirc) = 10$$

$$(\bigcirc - (\bigcirc - \bigcirc)\times(\bigcirc - \bigcirc))\times(\bigcirc - \bigcirc \times \bigcirc) = 10$$

$$(\bigcirc \times \bigcirc - \bigcirc)\div(\bigcirc + (\bigcirc + \bigcirc)\div(\bigcirc - \bigcirc)) = 10$$

Reach the target number by filling in the blanks in each equation.

A completed equation must incorporate each and every digit from 0 to 9.

You cannot move the operations (+, −, ×, ÷) found between blanks.

Order of operations applies!

10

You call *that* pressure?

In another of those 'wow, it's incredible we can do that!' moments, researchers have for the first time calculated the amount of pressure inside a single proton.

A proton is made up of 3 quarks, connected together by something that physicists call 'the strong nuclear force'. Quarks feel the strong force, yet protons which are made up of quarks, do not.

In mid-2018, a team headed by nuclear physicist Volker Burkert (who brings the noise at the Thomas Jefferson National Accelerator Facility – an American particle accelerator that is also known more cooly as 'The JLab') announced that the quarks within a proton are under pressure of 100 decillion pascals at the centre of the particle. A decillion is 10^{33} so that total pressure is about 10 times the pressure at the heart of a neutron star.

And even though a proton is incredibly small, it seems that the pressure is greatest at the centre and drops off as you move towards the edge.

This is the first time we've been able to look inside a subatomic particle and measure the forces at work.

Now, when I say 'we', of course I had nothing to do with it. But I am in complete awe of Volker and the team that did.

10

The *Good Will Hunting* Problem

One of the most famous mathematical problems in pop culture occurs in the movie *Good Will Hunting*. In fact the challenge, written up by a lecturer on a chalkboard in the corridor of the maths department, gained such widespread non-maths-world fame through the Academy Award-winning film, it is often called the '*Good Will Hunting* Problem'.

The challenge: 'Draw all the homeomorphically irreducible trees with $n = 10$.'

This isn't the easiest thing to understand off the top of your head, but as we do in these books, let's walk it through.

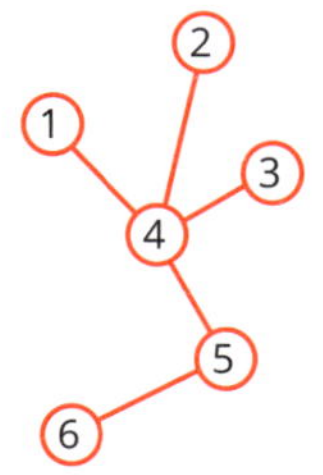

In mathematics, a 'tree' is simply a group of points, or 'nodes' joined by lines, with only one path between any two nodes. The first graph on the left here would be a tree with 6 nodes.

The second graph on the left would not be a tree, though, because the loop involving nodes 0, 1 and 2 means there are two different ways to get from 0 to 2 — you can go from 0 directly to 2 or from 0 to 1 to 2.

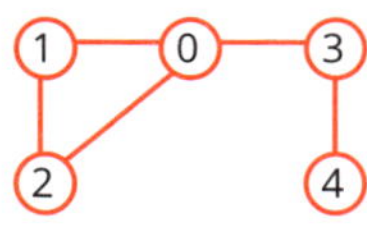

The 'with $n = 10$' bit is easy — we just need 10 nodes.

But the really tasty bit in the question, to the non-mathematical eye, is that the trees have to be 'homeomorphically irreducible'. This isn't as bad as it sounds. It just means that no node has exactly two lines attached to it. That is, every node in a path is a start point, an end point, or a junction of three or more lines.

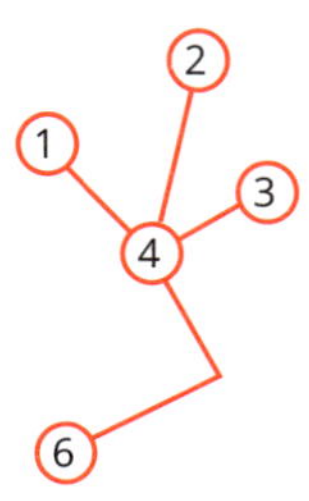

So while our first tree on the left is, well, a tree, it's not 'homeomorphically irreducible' because we could remove the node labelled 5 and the new tree (like the third graph on the left) would be essentially the same as far as we are concerned.

The tricky thing when you are trying to draw all these trees is that it is sometimes difficult to tell that two trees are actually the same.

For example, the first two trees of degree 8 on the right are actually the same.

You can see that, if you simply spin these branches around this way ...

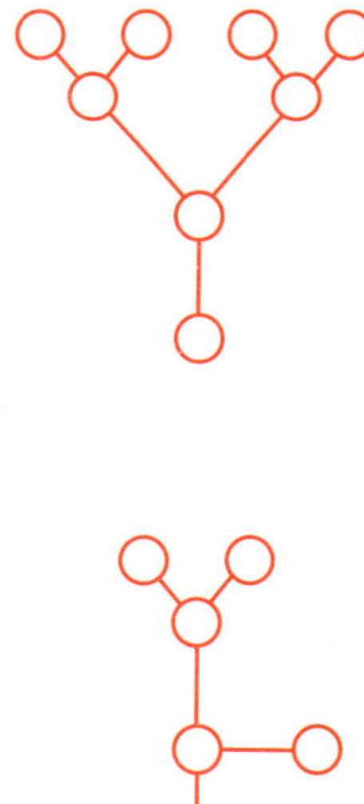

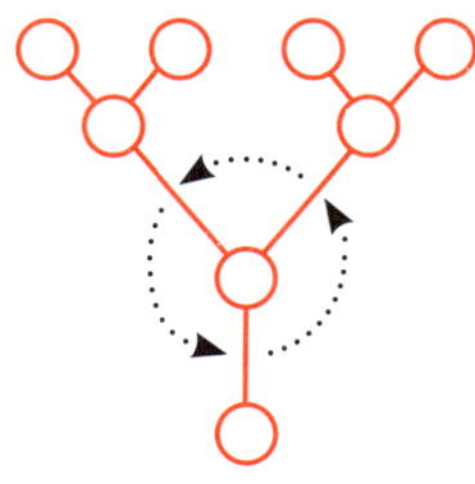

Got it? Care to have a shot at the '*Good Will Hunting* Problem'?

Having said it wasn't that easy to read at first, in the film the professor suggests it took the entire faculty 2 years to work out this problem.

Even by Hollywood standards this is an outrageously inflated claim. Have a crack yourself and see if you can improve on this '2 years for a whole faculty' benchmark.

You're trying to find 10 different trees.

Maybe you'd like to start by trying to find the '4 homeomorphically irreducible trees of order 8'.

Good luck.

9

Nine after nine

If you subtract a power of 10 from another power of 10, and then subtract 1, you get a certain pattern.

For example, $10^5 - 10^3 - 1 = 100{,}000 - 1000 - 1 = 98{,}999$ and:

$$10^7 - 10^2 - 1 = 10{,}000{,}000 - 100 - 1 = 9{,}999{,}899 \text{ and so on.}$$

You should be able to see that for any term $10^m - 10^n - 1$, the pattern is that you get a list of 9s interrupted by an 8. Also there are n 9s after the 8, or the 8 occurs in the $(m - n)$th position. And, of course, the final answer is m digits long.

'Okay Adam, even by your standards, this is getting a bit tedious ... where's the payoff here?' I hear you ask.

Well, bang.

Turns out the number $10^{6400} - 10^{6352} - 1$, that's 6399 nines interrupted by an 8 in the 48th position ... is prime.

If you're thinking of getting this as a neck tatt, you'll need something to take into the shop, so here you go. It's small, but I bunged it on the left. You get the drift.

But please ... don't get the neck tatt.

Sean's Syndesis

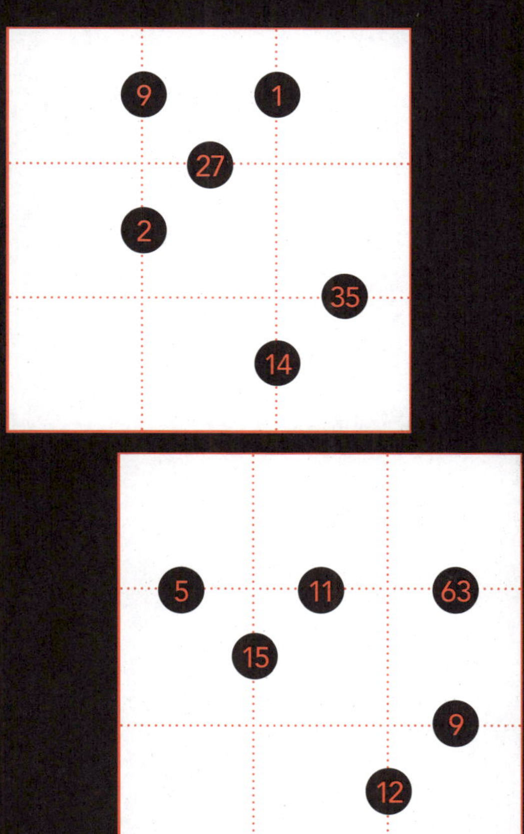

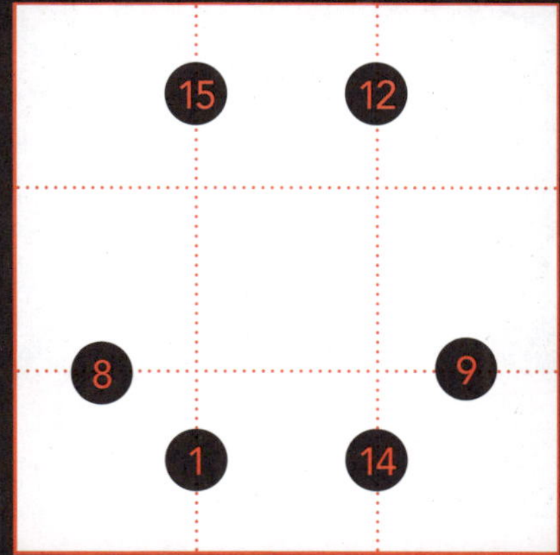

Enter the numbers 1 to 9 into each 3 × 3 grid so that each circled number is the result of adding, subtracting, multiplying, or dividing the two numbers in the cells it touches.

Head on back over to number 99 to read more about these awesome puzzles fresh from the mind of Sean Gardiner!

8

If you closely examined the first 1,000,000,000,000 digits of the decimal expansion of pi ...

(I know, it's on my to-do list too) you'd find that the number that occurs the most is 8, which pops up 100,000,791,469 times, just tipping out 2 and 4.

Played, number 8.

Jigglin' all the way

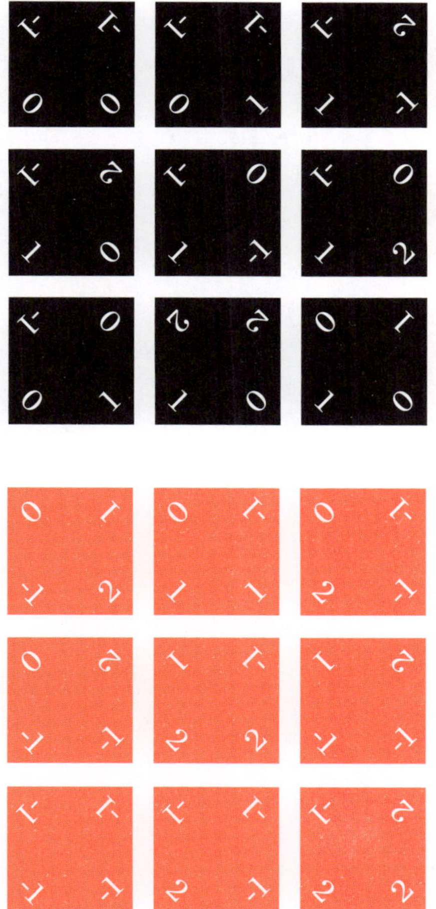

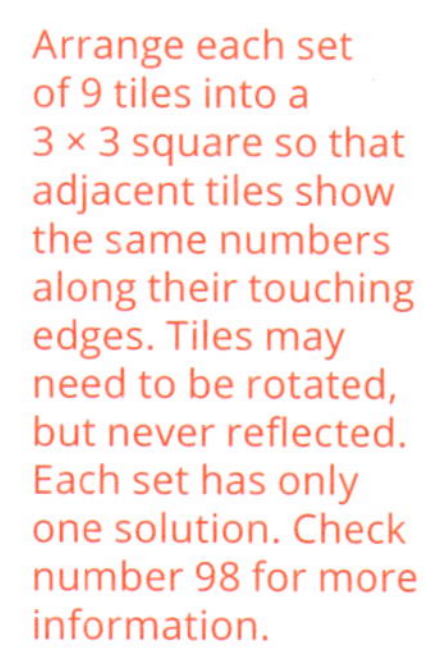

Arrange each set of 9 tiles into a 3 × 3 square so that adjacent tiles show the same numbers along their touching edges. Tiles may need to be rotated, but never reflected. Each set has only one solution. Check number 98 for more information.

For a downloadable cutout, head over to adamspencer.com.au/resources.

7

The (suit)case of the missing passports

Seven days before travelling overseas, you're all packed. Naturally you're feeling pretty chuffed about this impressive achievement, until you suddenly realise your kids have packed all the passports inside one of the bags.

You have 5 suitcases in a row.

Despite knowing exactly where they packed the passports, the kids seize the opportunity to tease you with a variation of a fiendish *Popular Mechanics* 'Riddle of the Week' that will neatly fit into your next book!

They tell you that each night, they will move all the passports one suitcase to the left or right, so that they are stashed in an adjacent suitcase every day. Each morning, you can look in one suitcase, and one suitcase only, to find the missing passports.

Aside from threatening to ground the kids for the rest of their lives, how can you guarantee you'll find the passports before you have to fly out?

Note, of course, you could just dive on in and start at suitcase number 1, working through to 5. But the fact that the passports move means you may miss them.

Likewise, there's little point checking the same suitcase over and over, since the kids could move the passports back and forth between only two suitcases. Dastardly!

New high score!

Welcome to our final game of High Scoring Equation where you fill in the blanks to win ~~real cash prizes~~ fame and glory! How'd you go so far? Brace yourself, cause this is the toughest one yet (and it's clue-free!). Go for it!

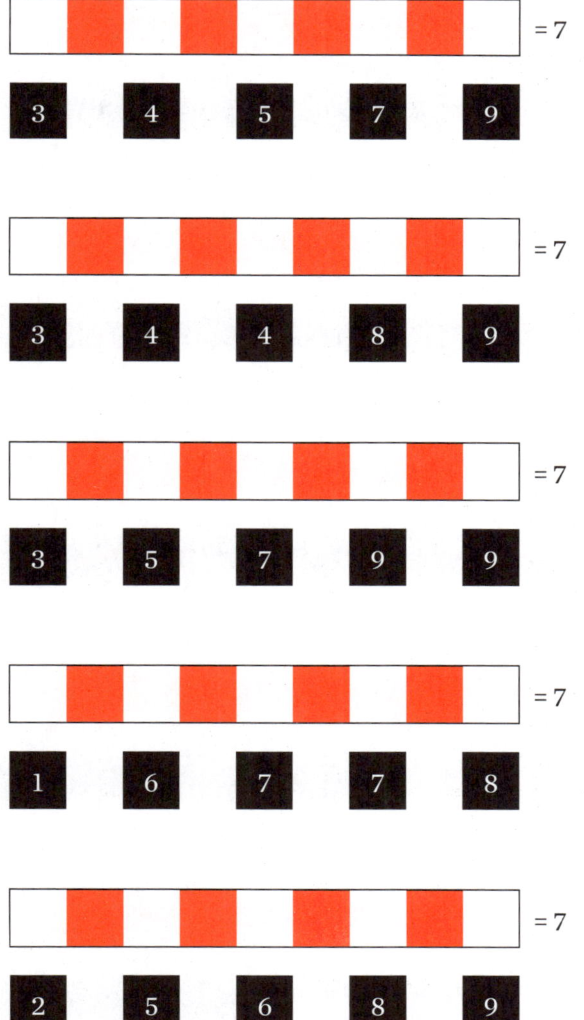

Reach the goal number by creating an equation using the provided numbers and your own choice of operators.

- The numbers should be placed in the white squares.
- Operations (+ – × ÷) are placed in orange squares.
- Order of operations matters, and you can't use brackets!
- Read the numbers off in order for your final score.
- Your goal is to find the equation that gives the highest score.

Head back over to number 97 if you need a refresher on these, otherwise ... get cracking!

6

During the first 6 seasons ...

of the TV series *Game of Thrones*, 1243 characters were killed.

That's an average of 21 deaths per episode.

More hexes!

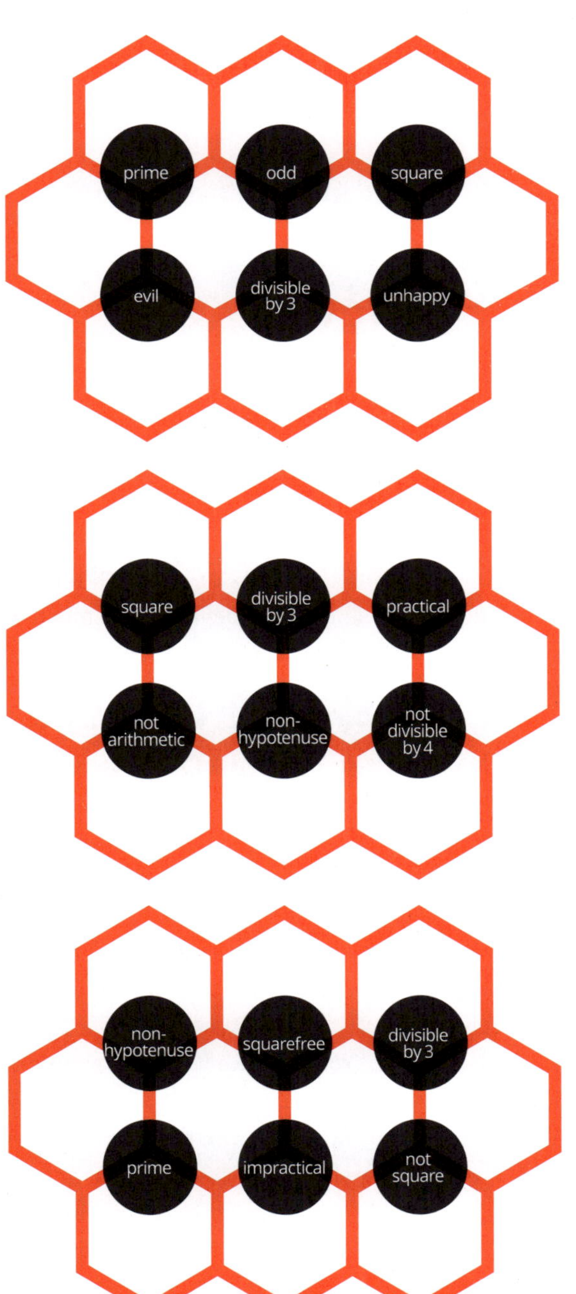

Place the numbers 11 to 20 into each hexagonal grid to satisfy all the rules.

Each cell is touching one or more categories in circles. The number in each cell must belong to all the categories touching it.

Need a hint? If the definitions are a bit tough for you, there's a full list of what these words mean at the end of the book.

5

You ~~lucky~~ funny duck!

I always preempt any piece I write about gambling with a warning.

I don't wish to sound paternalistic, but I think it's important that young people in particular are reminded as often as possible that some people ruin their lives by gambling.

In an age where sports gambling advertisements are so prevalent that kids often refer to their favourite team's chances on the weekend in terms of what a winning bet will pay, we have to tread carefully here.

If you read about someone winning a massive bet, please don't start gambling yourself assuming you will do the same and please, please, *please* only gamble any amount of money that you can happily lose without significant consequences.

Okay, I'll climb down from my pulpit now to mention that in 2018, Texas woman Margaret Reid bet $18 on a 'Pick 5' wherein she chose the potential winners of 5 horse races in a row.

Limousine Liberal, Maraud, Funny Duck and Yoshida all duly won and when Justify galloped to a win at the Kentucky Derby, Margaret Reid also won ... $1,200,000.

The Kentucky Derby is run annually in ... wait for it ... Louisville, Kentucky.

The race is 10 furlongs (2012 metres) long and the winner takes home only marginally more that our friend Margaret Reid: $1,425,000.

On the tiles

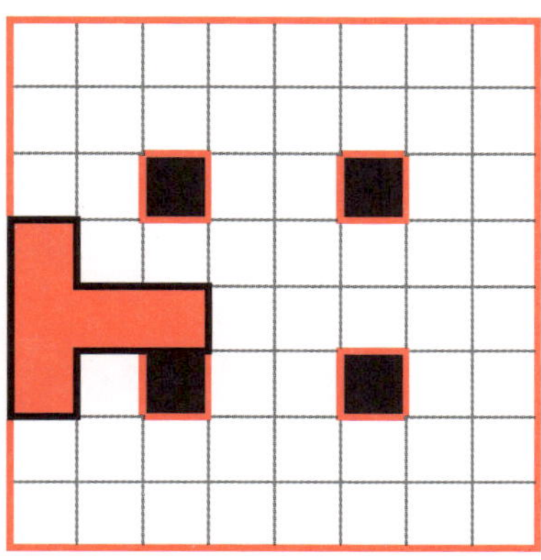

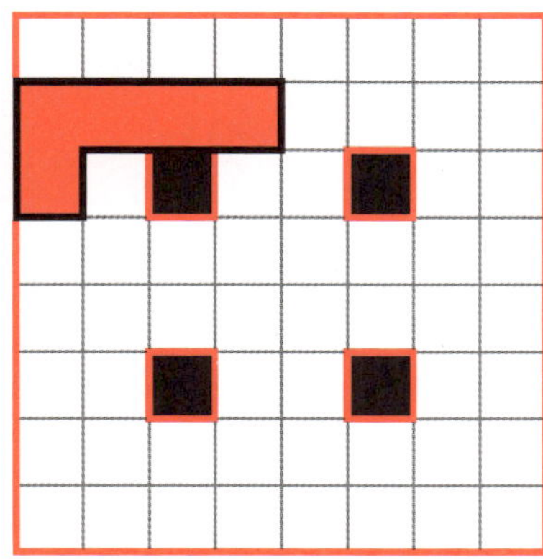

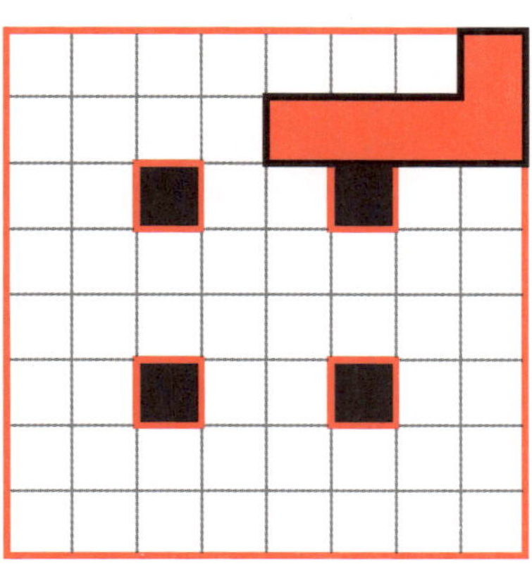

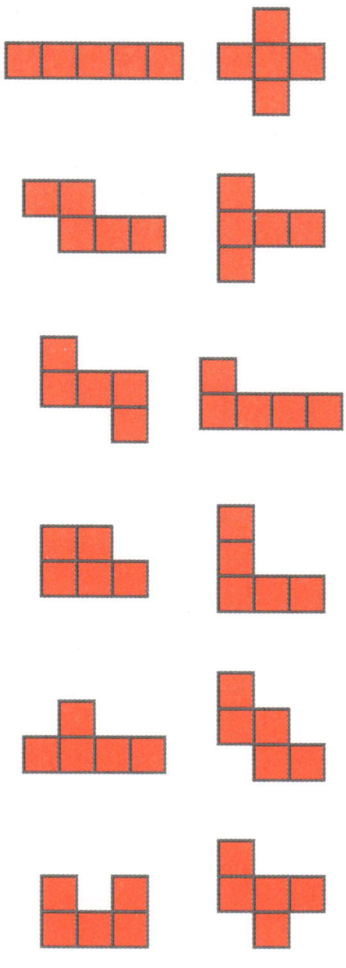

In each of these grids, finish tiling it so that it contains one of each pentomino type.

The 12 different pentominoes are shown here.

You may need to rotate and/or even flip some of them to get them to fit.

Head to adamspencer.com.au/resources if you'd like to download a copy of the puzzle to cut up!

4

Beat the gridlock!

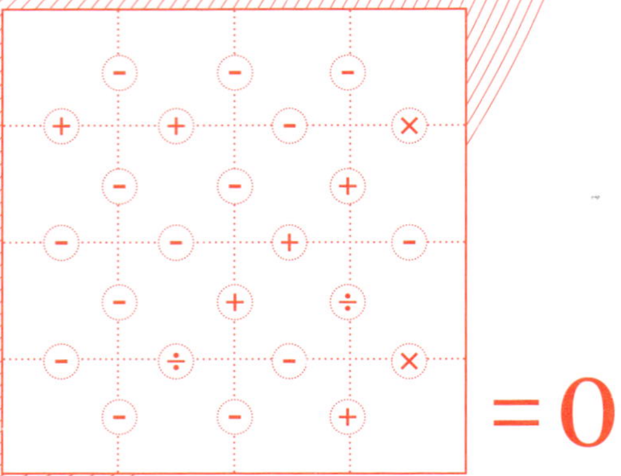

= 0

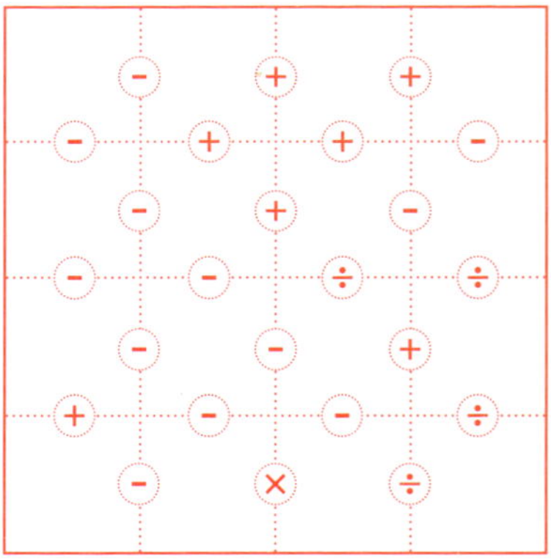

0 =

	−		+		+	
−		+		+		−
	−		+		−	
−		−		÷		÷
	−		−		+	
+		−		−		÷
	−		×		÷	

The rules for these are easy. The answers, well, they're a little tougher!

For each 4 × 4 grid, enter the numbers 1 to 16 so that each row and column equals the target number. As always, order of operations matters.

Can you break the gridlock?

If you don't like defacing this book, go to adamspencer.com/resources for a template to fill in.

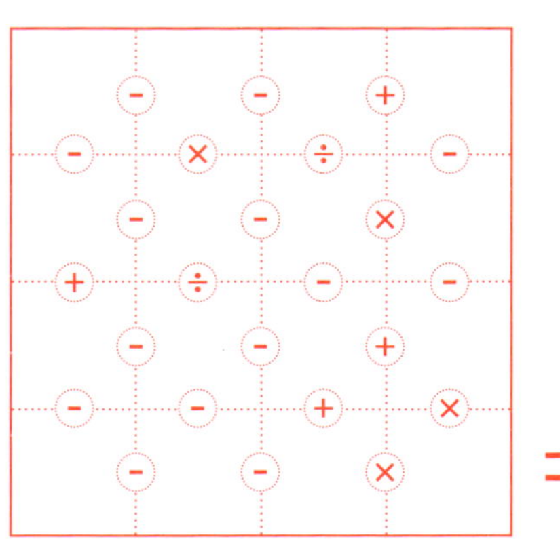

= 0

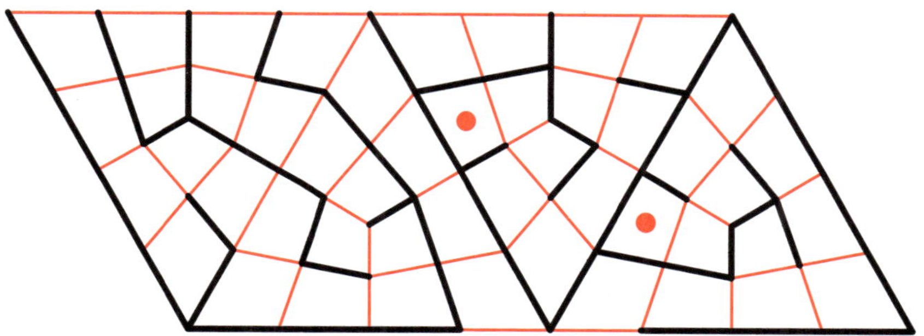

A maze

Do you like mazes?

At first glance this one above might appear pretty straightforward. Just get from one orange dot to the other.

University of Ljubljana (Slovenia) electrical engineer Izidor Hafner devised this maze ... with a difference. It's designed on a flattened tetrahedron.

A tetrahedron, remember, is a triangular pyramid — a 3D shape with 4 triangular faces.

So to complete this maze, you'll need to fold up the figure ... in your head.

Grab your pencil and ... go!

3

Prime, mate

If consecutive odd numbers are both prime, we call them twin primes. Three is part of the very first pair of twin primes, namely (3,5). The next are (5,7), followed by (11,13), (17,19), (29,31) and so on.

The only number that occurs in two pairs of twin primes is 5, because once primes get larger than 5, if you have 3 odd numbers in a row they can't all be prime — one of them must be divisible by 3.

If a prime number is not part of a twin pair it is called an isolated or single prime. It turns out that the overwhelming majority of prime numbers are isolated primes. This is a great example of something that isn't obvious in the world of small numbers where most of us spend most of our time, but which becomes apparent once you step back a bit and look at much larger numbers.

The largest twin primes we've found so far are $2996863034895 \times 2^{1290000} \pm 1$, with a whopping 388,342 decimal digits. The pair was discovered in September 2016.

So the question that may be bubbling up in your minds is this: 'Do the twin primes go on forever? Are there an infinity of twin primes?'

Well, good on you for thinking that — even if perhaps you needed some prodding! This question is called the 'twin prime conjecture'. Like many things about prime numbers, it's not a difficult question to understand, but in the almost 200 years that it had been considered, it had proven stubborn ... until ...

Hey, if you've got 17 minutes and want to listen to a hilarious and informative talk about the history of mega-prime numbers, delivered by an exquisitely handsome and very deluded Australian maths nerd, search 'Adam Spencer TED Talk' on YouTube and knock yourself out.

Enter Yitang Zhang (yet another really coolly-named mathematician) who in 2013 made a massive step forward. For someone who was almost completely unknown in the world of mathematics up to this point, he dropped a bomb when he proved that for some integer N less than 70,000,000, there are an infinity of pairs of primes that differ by N.

Now, this is not saying there are infinitely many twin primes, and 70,000,000 is a lot farther apart for a pair of primes to be than just 2, but it was a breakthrough to prove that there was any such structure to the prime numbers at all.

As often happens in cutting-edge mathematics, after years of seemingly going nowhere, once a trailblazer makes the initial breakthrough, other geniuses pile on in and help make improvements to their already brilliant work.

Among others, the great Australian mathematician Terence Tao (of course, who else!) got to work and, using improved methods over those used by Zhang, the gap has been reduced to 246. This is still a long way from solving the twin prime conjecture, but my man Yitang Zhang, you have done some *seriously* heavy lifting.

If there is only a finite list of twin primes then we could add them all up and get a finite sum. But in 1919 a mathematician by the name if Viggo Brun (I think I'll stop bothering pointing out all the awesome names here ... I mean – Viggo Brun!) ... showed that even if the sequence of twin primes is infinite, the sum of their reciprocals

$$\sum_p \left(\frac{1}{p}+\frac{1}{p+2}\right)=\left(\frac{1}{3}+\frac{1}{5}\right)+\left(\frac{1}{5}+\frac{1}{7}\right)+\left(\frac{1}{11}+\frac{1}{13}\right)+\cdots$$

converges to a finite number, which we call Brun's constant. We don't know its exact value yet because we don't even know if the list is infinite, but in 2002 two dudes named Pascal Sebah and Patrick Demichel summed the reciprio-cals of all the twin primes up to 10,000,000,000,000,000 to get an estimate of 1.902160583104.

While this may seem like a particularly arcane thing to spend your time on, Brun's mathematics was an important way of analysing primes at the time. Word, Viggo.

If you're totally maxing out on twin prime trivia, but would like a little bit more, go back and refresh your mind at 80 and 22.92067 about infinite sums with finite values.

3

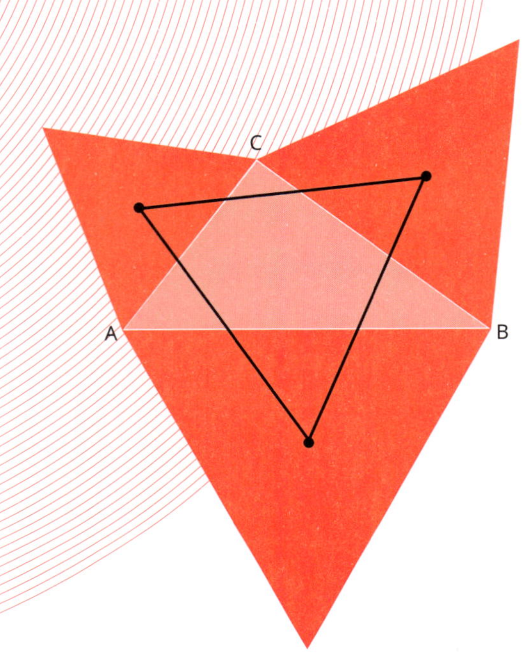

I've got a Bonaparte to pick with you

Joséphine de Beauharnais, Napoleon's first wife, was 32 when she married him. He was on the younger side of 26.

Nothing wrong with that of course, but it didn't matter much anyway since she reputedly knocked about 4 years off her age on the marriage certificate.

Napoleon added 18 months to his and, hey presto ... they were about the same age.

Take a triangle of any size, ABC.

Off the sides, form 3 equilateral triangles — these are triangles with all 3 sides having the same length. You know the drill.

Napoleon's Theorem states that the triangle formed by joining the centres of these equilateral triangles is itself equilateral.

While called Napoleon's Theorem, it is not clear if it was discovered by the French military supremo and one-time Emperor of France.

Regardless, it's a pretty cool little piece of geometry.

Drill bits (day 10!)...

Our final drill will be taken by #3, namely Jarrad McVeigh, the Sydney Swans superstar play-maker who, the day before sending this book to print broke his collarbone at the SCG! Rest up champion and I desperately hope to see you back on the field soon.

In which order did the players stand around the circle?

- #3 stood at the front of the circle.
- There were 4 people between #63 and #83.
- There was 1 person between #33 and #3.
- #33, #63, and #93 stood together in some order.
- From #43's perspective, #33 was on the left side of the circle.
- #53 stood closer to #43 than to #83.
- #3, #23, and #63 stood together in some order.
- From #93's perspective, #13 was on the right side of the circle.

Ten players stood in a circle for their daily drill.

They are numbered 3, 13, 23, 33, 43, 53, 63, 73, 83 and 93.

Each day a different player is the drill leader, which by delightful coincidence, corresponds to the chapter number and 'front position'.

Your task, coach, is to work out the order they stood in the circle each day given the list of clues provided by these fun-loving sporty puzzlers.

Game on!

3

Infinite. Nested. Radical.

We've spoken already of the genius of Ramanujan. You know, the Indian mathematician who died tragically young at just 32. His life was acclaimed in the movie *The Man Who Knew Infinity*. Sound familiar? He's the guy I mentioned earlier at number 10 to whom beautiful mathematics came in dreams.

Between 1911 and 1919 the big R-man sent a series of 58 problems to the *Journal of the Indian Mathematical Society*. One of the gnarliest questions was to find the value of the 'infinite, nested radical':

$$\sqrt{1+2\sqrt{1+3\sqrt{1+4\sqrt{1+\ldots}}}}$$

That sounds weird, but let's work through term by term. It is 'infinite' because the expression goes on forever; 'nested' because the square root signs are nested underneath each other; and 'radical' here refers not to revolutionary left-wingers who wish to overthrow society, but simply to terms that appear under a square root sign.

It probably won't come as a surprise to know that after 6 months no one had replied, so Ramanujan had to answer it himself.

Turns out:

$$\sqrt{1+2\sqrt{1+3\sqrt{1+4\sqrt{1+\ldots}}}} = 3$$

If that's not enough for you, if you yearn for more gorgeous infinite nested radicals, this is your lucky day.

Turns out that:

$$\sqrt{6+2\sqrt{7+3\sqrt{8+\ldots}}}=4$$

And if you're still not satisfied and you'd like one final infinite nested radical where you can actually follow how we get the value ... well, okay, you've asked nicely.

What is the value of x below?

$$x=\sqrt{2+\sqrt{2+\sqrt{2+\sqrt{2+\ldots}}}}$$

If you look closely you can see that the expression for x repeats itself under the first square root sign. So without calculating x we can see that

$$x=\sqrt{2+x}$$

And it is easy to see that this holds for $x = 2$.

So,

$$x=\sqrt{2+\sqrt{2+\sqrt{2+\sqrt{2+\ldots}}}}=2$$

Dude, that is (infintely nested) radical stuff!

2

There are only two references to sneezing in the Bible

But they are both doozies.

Depending upon your favourite translation they go something a little like this. (I should note I'm referring to the Christian Bible here.)

'Then he returned and walked in the house once back and forth, and went up and stretched himself on him; and the lad sneezed seven times and the lad opened his eyes.'
2 Kings 4:35

Which doesn't seem that interesting, but the lad who sneezed 7 times had supposedly been dead a short time earlier, so as far as sneezes go, that's pretty impressive.

And:

'His sneezes flash forth light, And his eyes are like the eyelids of the morning.'

Describing the Leviathan which as well as sneezing pure light has fearsome teeth and breath that melts coal.
Job 41:18

Nothing to be sneezed at.

Snakes alive!

→

7	+	5	–	2
+	3	–	8	×
5	+	9	×	5
–	9	×	6	=
9	×	8	=	2

→

6	×	3	×	5
–	4	×	8	+
3	×	1	+	6
+	7	–	4	=
1	×	9	=	2

→

8	×	3	–	8
×	6	×	2	+
9	×	1	+	1
×	1	–	3	=
8	+	4	=	2

In these grids, you can make paths from the top-left corner to the bottom-right by moving between adjacent squares.

You can move up, down, left or right, but you can never visit the same square twice.

And you guessed it, the paths must trace out the correct equation to reach the target number in the bottom right-hand corner.

For each grid, there are 3 different paths that all trace out correct equations.

Head on back to number 92 (if you haven't been there already) if you need a refresher.

2

Sublime prime time

Two is the smallest prime number.

If you're thinking, 'Hey, what about 1?', that's understandable.

For a long time the convention was to include 1 as a prime number. But there are many reasons why 1 is not considered prime. Here's a cool way of looking at it.

A piece of mathematics called 'the Fundamental Theorem of Arithmetic' (whoa okay — bringing out the big guns) says that every whole number greater than 1 is either prime itself or can be written uniquely as a product of primes.

This is a fancy way of saying that you can decompose 24 into prime factors $2 \times 2 \times 2 \times 3$ but there is no other combination of prime numbers that will give you 24. This is an important fact that is used in lots of higher mathematics.

If 1 was a prime number I could write:

$$24 = 2 \times 2 \times 2 \times 3 = 1 \times 2 \times 2 \times 2 \times 3 = 1 \times 1 \times 1 \times 1 \times 1 \times 2 \times 2 \times 2 \times 3$$

... and so on in an infinite number of ways.

The FTA would collapse and I don't know about you, but I certainly don't want to live in that sort of universe!

Here's another way to think about it.

When we multiply two prime numbers we must get an answer that is not prime.

If 1 and 3 are both prime then $1 \times 3 = 3$ is not prime because it is the product of two prime numbers.

But didn't you just say 3 was prime? So now I'm living in

a universe where 3 is both prime and not prime. Again, not for me, buddy.

As we mentioned back at number 59, my main man Euclid proved waaay back when that there are an infinite number of primes. They never stop. So there must be, at any time, a largest prime number that we know about.

In December 2017, a new chapter in mega-primes was written when a nifty distributed computing project called the Great Internet Mersenne Prime Search (GIMPS) proved that ... wait for it ... the 23,249,425-digit brute

$$2^{77,232,917} - 1$$

... is prime. The lucky volunteer on whose computer this massive number was crunched belonged to one Jonathan Pace. Cheers, JP!

Hold it, Adam – did you say *volunteer*? Indeed! At the moment, there are many distributed computing projects taking place online.

If you've got some spare grunt on a computer or PlayStation you've got lying around the house, it's not too hard to download the free software and, while you're out at school or work, you too could be searching for the next monster prime number, or gigantic prime arithmetic progression, or for that matter anything else which, if found, could see you preserved forever in the history of mathematics!

What's that? You're trying to remember where you can find that hilarious and informative 17-minute talk about the history of mega-prime numbers, delivered by an exquisitely handsome and very deluded Australian maths nerd which I mentioned only a page ago?

Say no more. Search 'Adam Spencer TED Talk' on YouTube and go crazy.

1

Doubly-true (still!) alphametics

Well, here we are at the very end of our countdown. Fittingly, for a book that I hope has amused, educated and at times *pushed* you too, we sign off with another alphametic.

These have been among the most challenging things we have tackled in this book so if you crack this last one, give yourself a massive pat on the back ... or like yourself on Insta ... or hit me up on Twitter (@adambspencer) ... or praise yourself in whichever way you see fit.

In honour of the number 1 this features not one, not two but *three* ones. If this seems like a lot, I'll let you in on a secret. The longest doubly-true alphametic we know of is chock full of ones ... 877 of them to be precise. It's in the sidebar for your enjoyment!

If you want to tackle that on paper ... even I would urge you to take a deep breath and maybe reconsider your options. Then again, who am I to stop you?

Anyway, to your final task in this entire hit list ... solve the doubly-true alphametic:

```
  NINETEEN
  THIRTEEN
     THREE
       TWO
       TWO
       ONE
       ONE
+      ONE
----------
  FORTYTWO
```

NINETEEN + NINETEEN + TEN + TEN + TEN + TEN + NINE + NINE + NINE + NINE + NINE + ONE [877 times] = THOUSAND

In praise of 1

A few pages back I referred to pi as perhaps the most amazing of all the numbers.

However, in closing, I'd like to pause and think for a little while about a number that probably doesn't get the love it deserves.

Whenever we talk about 'amazing numbers', the hit parade regularly includes pi and *i* (the square root of –1). Often infinity (which isn't a number), or zero (which didn't exist in many cultures for centuries and caused philosophers all manner of grief) gets a run. Then there's √2 (over which some ancient mathematicians were said to have committed murder to keep it a secret). Even the old 1 followed by a hundred zeroes – a googol – gets props. And deservedly so, they are all incredible. But rarely do we pause to give praise to 1.

In many ways it is the first number.

If you want to teach a kid to count, start at 1. But who knows? Perhaps the number 2 was actually named first because we took 1 so much for granted that only when you needed to specify that you didn't want 2 of something did we even give 1 a name. In that sense its history will forever be shrouded in secrecy. Nonetheless, I can tell you 1 rocks.

In the same way that multiplying prime numbers together makes them in some ways the building blocks of all numbers, when it comes to counting or adding, it all comes down to 1. 1 + 1 = 2 and from there it's a short 1-length step to 3, then 4, 5, 6 and you're away ... *forever*.

Of course, you could have started by turning left and stepping from zero to negative 1 then, the same distance away, negative 2 ... and if you've made it this far in the book you don't need me to tell you what happens next.

Multiply a number by 1 and that number stays the same – the 'multiplicative identity' for numbers is 1. Divide any number other than zero by itself? You get 1. Raise a number to the power of 0 ... you guessed it: 1.

The entire branch of complex numbers, which plays a crucial role in everything from electrical engineering, climate modelling and the science of flight, through to quantum mechanics and our understanding of the world on the smallest of scales ... essentially comes from asking what lies at heart of the difference between 1 and its negative.

Whether you're talking the number 1, the polynomial 1, the 3×3 identity matrix I, the function $f(x) = x$, the set {1}, the word 'true' in computer science or even the Boolean algebra gate:

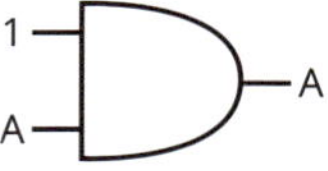

... the concept of 1 lies at the heart of *all* mathematics.

And in so many more fields. To be number 1 symbolises the triumph of ambition, one is our sense of self, falling in love is the moment when 'two become one', John Lennon yearned for a world that will 'be as one'.

So whether you pronounce '1' as one or une, eins, eitt, en, unu, uma, uno, odin, jeden, ikk, yek, hito, hana, nye, yī, qi, nueng, bir, ikte, egy, nitokska, wanggany, ataqan, chaffa, bat, youn, wāhad, ehad, meuj, besik, dua, onnu, onen, een, yan in the Shilha/Tachelhit dialect of southwestern Morocco or you prefer the Klingon wa', take a moment today to salute this most beautiful of numbers and our final destination in this *Top 100* ...

One, you are our number 1.

Answers

Congratulations, brave and hardy traveller!

You've ridden the *Top 100* countdown from 100 to 1 ... and what a ride it was! No doubt you've wrestled with many of the problems we threw at you. Probably with varying degrees of success, and even enjoyment. If you haven't snuck a peek already, here they are ... the answers.

But these answers are far more than just a guide to whether you give your work a tick or cross with the red pen of your mind. Hopefully they will walk you through *how* we arrive at the solution and, in the event that you couldn't crack that particular nut, give you the skills to be able to tackle it next time.

Some answers go through several stages. Feel free to read the first bit of the solution, then go back and, armed with this new knowledge, lock horns again with the thorny foe.

And two more important provisos. Firstly, there is often more than one way to skin the mathematical cat*. You may well have reached the same answer in a different, but equally valid way. Secondly, my refusal to pay heed to deadlines and to continue to cram more and more into this book often even after we have sent it to print (#loophole!) means there is every chance that a couple of 'typos' have 'gremlin-ed' their way into the text. If you've attempted a question 6 times and continually turned up the same answer which doesn't seem to match mine, hit me up at book@adamspencer.com.au and we can compare war stories.

I hope these answers give you more smiles than stress lines. Cheers!

* Skinning cats is totally gross and barbaric. Please take this reference as a mere figure of speech, not a motto by which to live your life.

100

Big, fat 100!

It should be obvious that $10 = 2 \times 5$. So in the massive equation 100!, every time a 5 multiplies with a 2, we get a 10 and no matter what else that multiplies with we keep the 0 from this 10 in our answer.

So $5 \times 8 = 40$ and $37 \times 5 \times 8 = 1480$, keeping the 0.

Firstly, rewrite the equation $100 \times 99 \times 98 \times 97 \times 96 \times 95 \times \ldots \times 4 \times 3 \times 2 \times 1$ by rewriting each number reduced to its basic factors:

> $100 = 2 \times 2 \times 5 \times 5$, $98 = 2 \times 7 \times 7$, $97 = 97$ (97 is called a prime number for this reason), $96 = 2 \times 2 \times 2 \times 2 \times 2 \times 3$, $95 = 5 \times 19$ and so on.
>
> So $100 \times 99 \times 98 \times 97 \times \ldots \times 4 \times 3 \times 2 \times 1 = 2 \times 2 \times 5 \times 5 \times 2 \times 7 \times 7 \times 97 \times 2 \times 2 \times 2 \times 2 \times 2 \times 3 \times 5 \times 19 \times \ldots \times 5 \times 2 \times 2 \times 3 \times 2 \times 1$.

But when we multiply a string of numbers together we can change the order and still get the same number. You can check that $2 \times 3 \times 13 \times 7 = 13 \times 7 \times 2 \times 3$.

Our factorisation of 100! will contain more 2s than 5s. So you should be able to see that by dragging out all the 5s and matching them up with 2s we can generate all the 10s that give us all the 0s in the tail of 100!. To know how long this tail is we just need to know how many 5s are in the factorisation.

The numbers that give us factors of 5 are 5, 10, 15, 20, 25 ... 95 and 100. There are 20 numbers on this list. But we get a bonus second factor of 5 from the numbers 25, 50, 75 and 100. So there should be 24 zeroes in the tail of 100!.

Indeed there are:

> 100! = 93326215443944152681699238856266700490715968264381621468592963895217599993229915608941463976156518286253697920827223758251185210916864000000000000000000000000

$$(8+2)\times(4+(9\times7+3)\div(6+5))=100$$

$$((5\times(7-3)+8)\div2+6)\times(9-4)=100$$

$$(7+9\div(5-2))\times(8-3)\times(6-4)=100$$

$$(6\times(4-(9+5)\div7)+8)\times(3+2)=100$$

$$((8-4)\times(9-6)+7)\times(3+2)+5=100$$

For the third example, there were two other solutions (thanks to intrepid readers Chris Fraser and Peter Profiris for spotting these ones):

$$(9+2\div(6-4))\times(8-3)\times(7-5)=100$$

$$(9+4\div(6-2))\times(8-3)\times(7-5)=100$$

99

4		5		1
		15		9
6		3		9
3		11		63
2	16	8		7

5		2		7
11		6		4
6	48	8		3
		9		7
9		1		4

1		4	1	3
		11		
2	9	7	13	6
8	3	5	45	9

98

4	3	3	3	3	3
3	2	2	2	2	4
3	2	2	2	2	4
2	1	1	4	4	3
2	1	1	4	4	3
4	4	4	2	2	3

1	2	2	4	4	2
3	1	1	4	4	1
3	1	1	4	4	1
2	3	3	4	4	1
2	3	3	4	4	1
3	3	3	4	4	4

1	2	2	1	1	4
3	4	4	3	3	3
3	4	4	3	3	3
2	2	2	3	3	3
2	2	2	3	3	3
1	3	3	3	3	4

97

Flight simulator

Question 1	Question 2	Question 3	Question 4
E	D	D	B

Each pattern is a simple geometric manipulation of a combination of elements.

In the first 3, for example, consider the path of the dot by itself and then look at how the remaining pieces are moving.

In the 4th example the triangle moves in relation to the square and eventually vanishes because its base is the same length as the square.

9	×	3	+	7	×	5	×	2	= 97
7	×	5	×	3	–	7	–	1	= 97
8	×	7	×	2	–	9	–	6	= 97
5	×	5	×	4	–	6	÷	2	= 97
9	×	6	×	5	÷	3	+	7	= 97

96

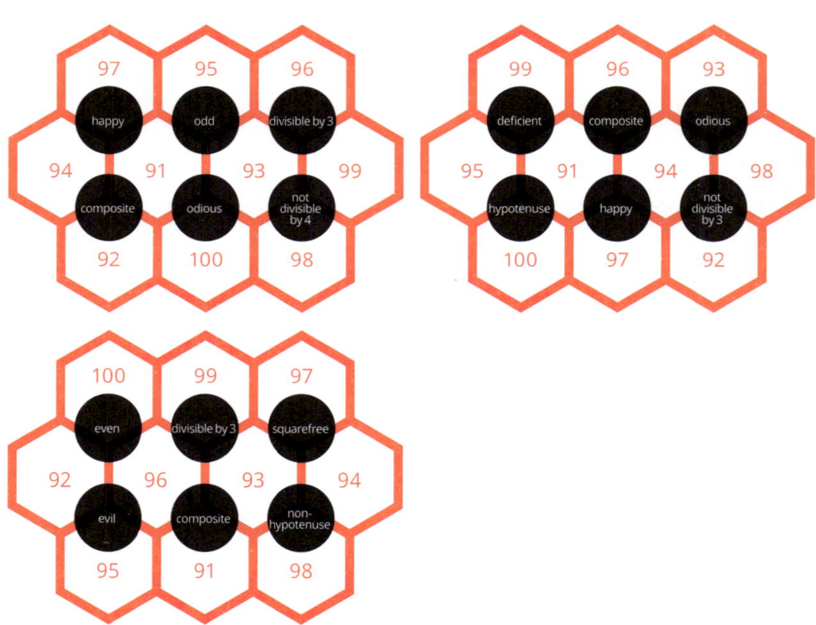

95

pq your interest

The other 3 semiprimes greater than 90 are 94, 93 and 91.

94

15	+	11	−	9	−	8
÷		+		−		×
3	+	7	+	13	−	14
−		−		+		÷
6	×	4	+	1	−	16
+		−		+		+
10	+	5	−	12	÷	2

8	+	9	−	13	+	5
×		−		+		×
3	÷	14	×	6	×	7
−		+		+		−
16	+	4	×	2	−	15
+		+		−		−
1	×	10	−	12	+	11

7	−	10	−	4	+	16
+		+		+		−
1	−	8	÷	2	+	12
×		+		−		+
13	+	5	+	6	−	15
−		−		+		÷
11	−	14	+	9	+	3

93

Going clockwise starting with the front position:
#93, #53, #43, #63, #73, #13, #33, #23, #83, #3

92

→

5	**×**	**9**	×	1
×	**2**	**×**	3	−
7	**−**	7	×	7
×	**1**	**+**	4	=
7	×	**3**	**=**	**92**

→

5	**×**	**9**	×	1
×	**2**	**×**	**3**	**−**
7	**−**	7	**×**	**7**
×	**1**	**+**	**4**	**=**
7	**×**	**3**	=	**92**

→

5	**×**	**9**	**×**	**1**
×	**2**	**×**	**3**	**−**
7	**−**	7	**×**	**7**
×	**1**	**+**	**4**	**=**
7	**×**	**3**	=	**92**

→				
7	×	8	×	7
−	2	×	2	−
1	×	9	×	2
+	6	−	5	=
1	+	6	=	92

→				
7	×	8	×	7
−	2	×	2	−
1	×	9	×	2
+	6	−	5	=
1	+	6	=	92

→				
7	×	8	×	7
−	2	×	2	−
1	×	9	×	2
+	6	−	5	=
1	+	6	=	92

→				
9	×	1	×	2
−	4	+	9	+
9	−	9	+	1
×	7	×	3	=
5	−	9	=	92

→				
9	×	1	×	2
−	4	+	9	+
9	−	9	+	1
×	7	×	3	=
5	−	9	=	92

→				
9	×	1	×	2
−	4	+	9	+
9	−	9	+	1
×	7	×	3	=
5	−	9	=	92

91

Substitute 828310483 for NINETYONE in the original equation. You should then be able to see the following: in the right-hand column (or 'units' column), we have all the digits already, but we realise we carry a 3 across to the tens column.

In the tens column we then get 3 + 8 + V + 3 + 3 + 3 + V = 8; or 20 + 2 × V = C8, a 2-digit number with first digit C and second digit 8. Solutions are V = 4 or V = 9, but we already have O = 4, so V = 9 and we carry the C = 3. So in the hundreds column we now have all digits, but we still work it out to see we are carrying a 2 into the thousands column. In the thousands column we have 2 + F + 3 + 1 + F = C0, where C now stands for the new number we will carry. This is 6 + 2 × F = C0, so F = 2 or F = 7. Again we already have I = 2, so F = 7 and we carry C = 2. In the ten thousands column we now have 2 + L + 3 + 0 = C1, so L = 6 and we carry C = 1. Keep going like this and you get R = 5 from the millions column and we are done.

So our answer is:

E = 3, F = 7, I = 2, L = 6, N = 8, O = 4, R = 5, T = 1, V = 9, Y = 0

```
          483
         7293
          138
       363938
     82831338
+   745107293
-------------
    828310483
```

90

$6 \times (7-2) \times ((8-4) \div (5-3)+1) = 90$

$((8-5) \times (7-4)+6) \times (3+2+1) = 90$

$(7-1) \times (3+2) \times (5-(8+4) \div 6) = 90$

$(6+4) \times ((7-3) \div 2+1) \times (8-5) = 90$

$(((4-1) \times 3+8) \times (6+5)-7) \div 2 = 90$

89

2		3		6
(8)				(1)
4		1		5
(28)		(7)		(45)
7		8	(17)	9

4		8	(8)	1
(13)		(56)		
9		7	(42)	6
(18)				
2	(3)	5		3

3	(7)	4	(5)	9
7	(35)	5		6
(8)				
1	(8)	8	(10)	2

88

3	2	2	1	1	3
4	1	1	3	3	2
4	1	1	3	3	2
3	2	2	2	2	3
3	2	2	2	2	3
3	3	3	4	4	3

1	3	3	1	1	4
1	1	1	4	4	4
1	1	1	4	4	4
3	2	2	3	3	4
3	2	2	3	3	4
2	1	1	3	3	1

1	1	1	2	2	2
1	3	3	3	3	2
1	3	3	3	3	2
2	2	2	4	4	2
2	2	2	4	4	2
4	3	3	3	3	1

87

9	×	5	+	7	×	3	×	2	= 87
9	+	7	×	4	×	3	–	6	= 87
9	×	6	+	8	×	5	–	7	= 87
8	×	6	×	4	÷	2	–	9	= 87
8	×	4	×	3	–	5	–	4	= 87

86

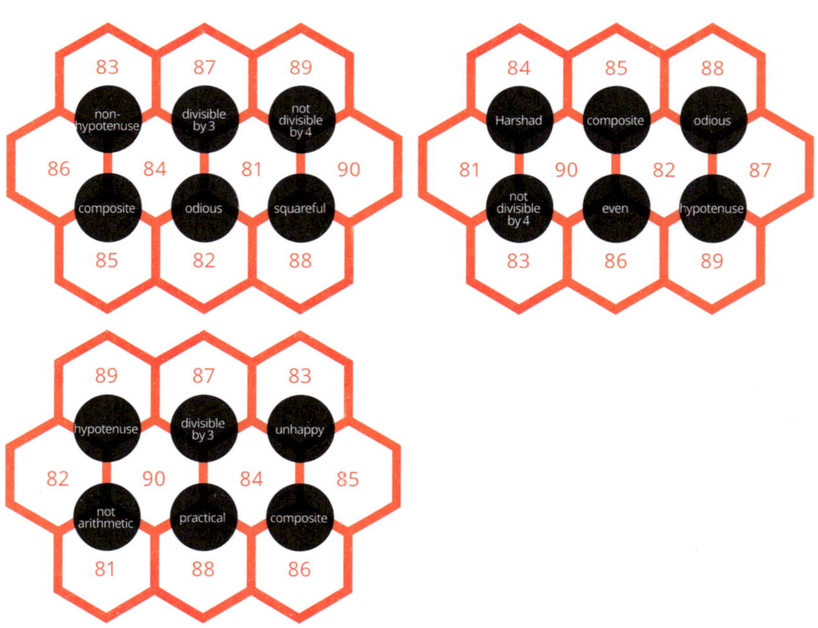

85

84

15	-	14	÷	6	×	3
+		÷		×		+
12	-	7	+	1	+	2
-		÷		-		+
9	-	4	+	11	-	8
-		×		+		-
10	+	16	-	13	-	5

1	×	10	-	11	+	9
×		+		+		-
7	-	2	×	6	+	13
÷		+		-		-
14	-	8	÷	4	×	3
×		-		-		+
16	+	12	-	5	-	15

15	+	16	-	10	-	13
+		-		-		-
8	-	7	-	5	+	12
-		×		+		+
14	×	2	-	9	-	11
-		+		÷		-
1	×	6	÷	3	×	4

83

The big swim

One of the critical things in any mathematics problem is working out what information is important and what can be ignored.

In this case, the volume of the pool is irrelevant because we are told how long it takes the pipes to fill the whole pool. The question instead is 'how do I compare the "24 hour" pipe with the "12 hour" and "8 hour" pipe'?

Well, how do you compare 8, 12 and 24 hour rates? They can all easily become 24 hour rates, can't they? So ask yourself: in 24 hours, how many pools could these pipes fill? Pipe C fills one pool every 24 hours. Pipe A fills two pools in that time and Pipe B would fill 3.

So working together, in 24 hours these pipes could fill 6 pools. It is suddenly obvious that together they can fill one pool in 4 hours.

Going clockwise starting with the front position:
#83, #3, #63, #13, #43, #93, #53, #73, #23, #33

82

→				
9	−	4	+	2
×	2	+	2	−
4	×	4	×	5
×	8	−	8	=
7	−	7	=	82

→				
9	−	4	+	2
×	2	+	2	−
4	×	4	×	5
×	8	−	8	=
7	−	7	=	82

→				
9	−	4	+	2
×	2	+	2	−
4	×	4	×	5
×	8	−	8	=
7	−	7	=	82

→				
5	×	5	×	1
+	8	+	8	+
4	+	2	−	8
−	2	+	7	=
8	×	4	=	82

→				
5	×	5	×	1
+	8	+	8	+
4	+	2	−	8
−	2	+	7	=
8	×	4	=	82

→				
5	×	5	×	1
+	8	+	8	+
4	+	2	−	8
−	2	+	7	=
8	×	4	=	82

→				
2	−	5	×	7
−	4	×	5	×
2	+	6	×	3
×	6	−	4	=
7	+	8	=	82

→				
2	−	5	×	7
−	4	×	5	×
2	+	6	×	3
×	6	−	4	=
7	+	8	=	82

→				
2	−	5	×	7
−	4	×	5	×
2	+	6	×	3
×	6	−	4	=
7	+	8	=	82

81

We already have values for 0, 1, 2, 3, 4 and 5 so we are looking for values of F, N, O and W with possibilities 6, 7, 8 and 9.

In the units column we have 4 + 4 + 2 + 2 = C2 where we carry the digit C into the tens column. So clearly we carry a 1 across. In the tens column we then get 1 + N + N + 3 + 3 = C3. The only possibility is N = 8, and we carry C = 2. In the hundreds column, we get 2 + 0 + 5 + 8 + F = C1. The letters O and F can only be 6, 7 or 9. They must be 7 and 9, but we are not sure which order; and we carry and C = 2. This means W = 6. The thousands column then gives us 2 + 8 + 4 + 5 = CO, so O = 9 leaving F = 7.

Substituting all this back to check shows we have found the solution:

E = 4, F = 7, G = 0, H = 1, I = 5, N = 8, O = 9, T = 3, W = 6, Y = 2

```
      984
     8584
   364832
+   75732
---------
   450132
```

80

(5 + (9 − 3)÷(6 − 4) + 2)×(7 + 1) = 80

((3 + 2)×(6 ÷ (9 − 7) + 1) − 4) × 5 = 80

(9 − 1)×(3 + 2)×(4 − (7 + 5) ÷ 6) = 80

((6 + 3) × 9 × (7 − 2) − 5)÷(4 + 1) = 80

(7 + 9 ÷ (2 + 1))×(6 × (5 − 3) − 4) = 80

79

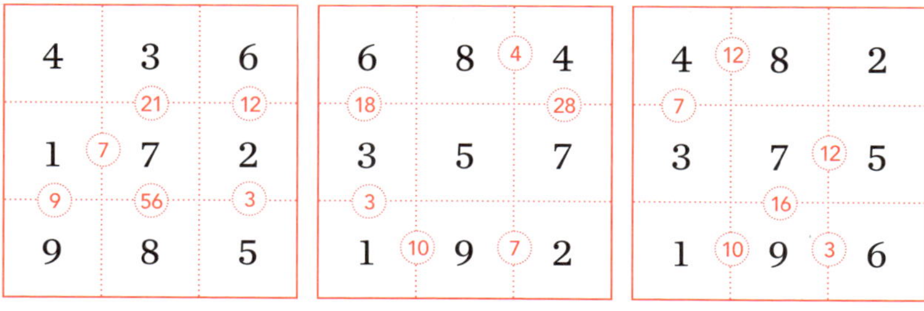

What are the chances?

The set of residents who drink flat whites comprises 79%, and the group who don't drink flat whites comprises 21%.

Likewise, non-latte drinkers = 49%, non-water drinkers = 10% and non-tea drinkers = 20%.

Since we know that every Numbervillean abstains from something, we can deduce that these 4 segments together comprise 100% of the residents.

Since the 4 segments total 100%, there is obviously no overlap whatsoever.

So every resident drinks 3 of the 4 beverages, and 100% drink coffee.

78

Good ol' 78s

1.33 seconds. Like our swimming pool question back at number 83, the distances 6 cm and 18 cm are distractions here. The record must rotate at the same rpm for the entire record — otherwise you'd get some weird warping phenomena! So a lap around each groove must take the same time no matter how far it is from the centre.

Prime-time magic

No, it's not possible. If we add all the numbers up, we get a total of 78. In a magic square, each row and column need to be a third of that or 26 in this case.

Problem is, only one of the primes is even, which means that in two of the rows and columns, we'd have to produce an even number by adding 3 odd numbers, which is impossible.

2	1	1	4	4	2
3	3	3	2	2	2
3	3	3	2	2	2
1	1	1	2	2	3
1	1	1	2	2	3
2	4	4	1	1	2

3	1	1	3	3	2
2	3	3	1	1	2
2	3	3	1	1	2
3	2	2	2	2	2
3	2	2	2	2	2
2	1	1	4	4	3

1	2	2	3	3	3
1	2	2	1	1	1
1	2	2	1	1	1
2	4	4	3	3	1
2	4	4	3	3	1
4	1	1	2	2	4

77

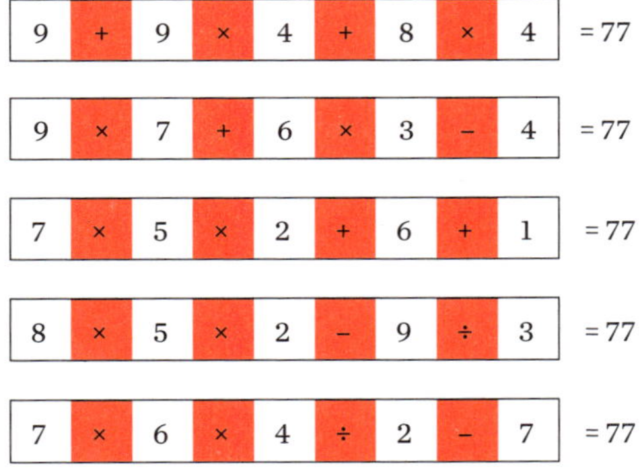

Take me out to the geek fest

$715 = 5 \times 11 \times 13$ and $5 + 11 + 13 = 2 + 3 + 7 + 17 = 29$. Further $77 = 7 \times 11$ and $78 = 2 \times 3 \times 13$ and $7 + 11 = 2 + 3 + 13 = 18$ making 77 and 78 a Ruth-Aaron pair.

And finally (and real props if you got this):

$89460294 = 2 \times 3 \times 7 \times 11 \times 23 \times 8419$,
$89460295 = 5 \times 4201 \times 4259$,
$89460296 = 2 \times 2 \times 2 \times 31 \times 43 \times 8389$,
and $2 + 3 + 7 + 11 + 23 + 8419 = 5 + 4201 + 4259 = 2 + 31 + 43 + 8389 = 8465$.

76

78
73
75
not divisible by 4
prime
deficient
74
79
71
77
odious
non-hypotenuse
not divisible by 3
76
72
80
79
71
73
non-hypotenuse
unhappy
hypotenuse
77
72
80
75
composite
abundant
squareful
74
78
76
78
80
72
squarefree
unhappy
non-hypotenuse
74
73
71
77
impractical
prime
not divisible by 4
76
79
75

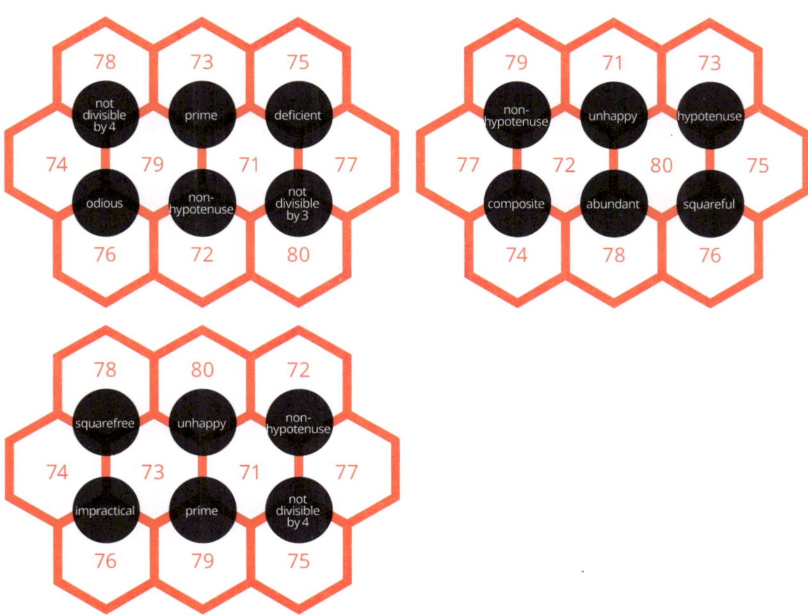

75

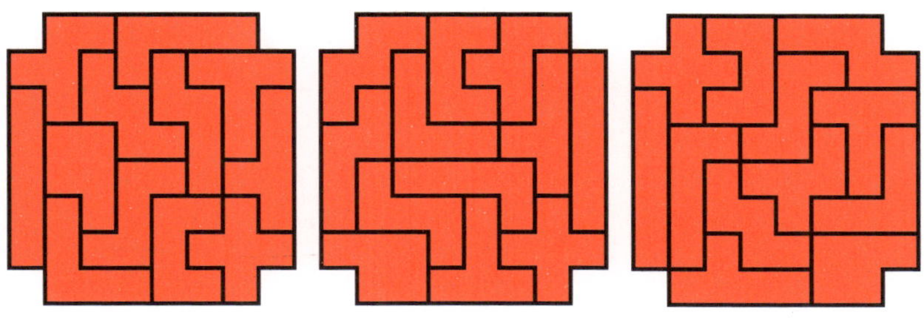

74

12	−	10	−	9	+	14
−		+		−		+
1	+	4	+	16	÷	8
−		−		+		−
15	+	13	−	7	×	3
+		+		×		×
11	+	6	−	2	×	5

5	−	14	+	13	+	3
÷		+		−		+
10	+	15	−	12	−	6
×		−		÷		+
16	÷	2	−	8	+	7
−		×		×		−
1	+	11	+	4	−	9

13	−	12	÷	6	×	3
+		−		+		+
8	−	16	+	15	×	1
÷		+		−		−
2	+	7	+	9	−	11
−		+		−		+
10	÷	4	÷	5	×	14

73

Going clockwise starting with the front position:
#73, #43, #83, #13, #23, #3, #53, #93, #63, #33

72

→

7	−	4	+	3
+	2	×	2	−
7	−	5	+	3
×	4	×	8	=
9	×	8	=	72

→

7	−	4	+	3
+	2	×	2	−
7	−	5	+	3
×	4	×	8	=
9	×	8	=	72

→

7	−	4	+	3
+	2	×	2	−
7	−	5	+	3
×	4	×	8	=
9	×	8	=	72

→

4	×	4	×	4
+	3	−	3	×
4	+	6	×	9
−	4	−	2	=
3	−	4	=	72

→

4	×	4	×	4
+	3	−	3	×
4	+	6	×	9
−	4	−	2	=
3	−	4	=	72

→

4	×	4	×	4
+	3	−	3	×
4	+	6	×	9
−	4	−	2	=
3	−	4	=	72

→

1	+	7	+	2
×	1	×	4	+
9	+	4	+	7
−	8	−	7	=
9	+	4	=	72

→

1	+	7	+	2
×	1	×	4	+
9	+	4	+	7
−	8	−	7	=
9	+	4	=	72

→

1	+	7	+	2
×	1	×	4	+
9	+	4	+	7
−	8	−	7	=
9	+	4	=	72

71

Looking at the two columns furthest to the left we can see that in the millions column we must be carrying a 1 across into the tens of millions column. So N = 8. Substituting that back into the units column gives us E = 6. From the tens column we get V = 2 and W = 5 follows easily soon after.

So our solution is:

E = 6, F = 3, I = 9, N = 7, S = 8, T = 4, V = 2, W = 5, X = 1, Y = 0

$$
\begin{array}{r}
3926 \\
891 \\
8914667 \\
79764667 \\
+\quad 456740 \\
\hline
89140891
\end{array}
$$

70

(6 + 1)×(9 − 4)×(8 ÷ (5 − 3) − 2) = 70

(6 + 4)×((9 + 3)×(5 + 1) ÷ 8 − 2) = 70

(6 ×((9 − 5) ÷ 2 + 1) − 8)×(4 + 3) = 70

(9 + 5)×(8 + 2)÷(4 × (3 − 1) − 6) = 70

((9 − 5)×(8 − 4) − 6)×(3 × 2 + 1) = 70

69

6	(4)	2	(1)	3
9	(16)	7	(6)	1
		(56)		
4		8	(3)	5

9		6		4
(17)				(28)
8	(6)	2		7
(8)				(10)
1		5	(15)	3

1	9	7
(9)	(36)	(42)
8	4	6
(6)	(1)	(1)
2	3	5

68

2	2	2	3	3	1
2	1	1	1	1	1
2	1	1	1	1	1
1	2	2	2	2	3
1	2	2	2	2	3
4	4	4	2	2	1

2	1	1	4	4	2
2	3	3	1	1	1
2	3	3	1	1	1
2	2	2	2	2	1
2	2	2	2	2	1
2	2	2	1	1	2

3	3	3	1	1	4
2	1	1	1	1	2
2	1	1	1	1	2
1	2	2	1	1	3
1	2	2	1	1	3
4	2	2	4	4	1

67

The big D

Writing our first example $24^5 + 28^5 + 67^5 = 3^5 + 54^5 + 62^5 = 1{,}375{,}298{,}099$ as (24, 28, 67) (3, 54, 62) 1,375,298,099, the other 4 solutions are:

(18, 44, 66) (13, 51, 64) 1,419,138,368
(21, 43, 74) (8, 62, 68) 2,370,099,168
(56, 67, 83) (53, 72, 81) 5,839,897,526
(49, 75, 107) (39, 92, 100) 16,681,039,431

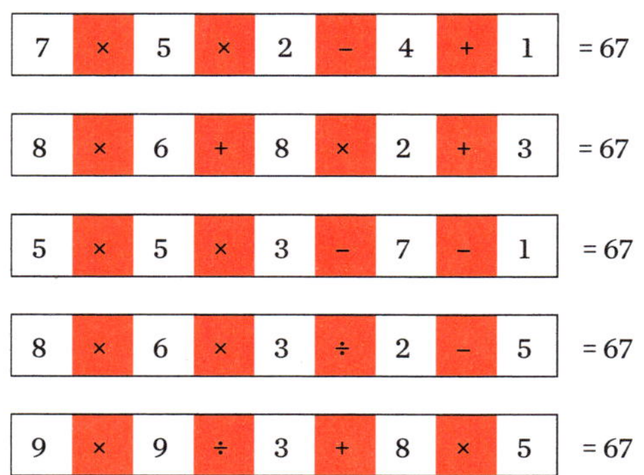

66

65
70
61
evil
composite
odd
68
63
69
67
squareful
divisible by 3
non-hypotenuse
64
66
62
67
64
61
impractical
squareful
hypotenuse
69
63
68
70
divisible by 3
evil
even
66
65
62
70
69
68
even
divisible by 3
squareful
62
66
63
64
non Harshad
evil
non-hypotenuse
61
65
67

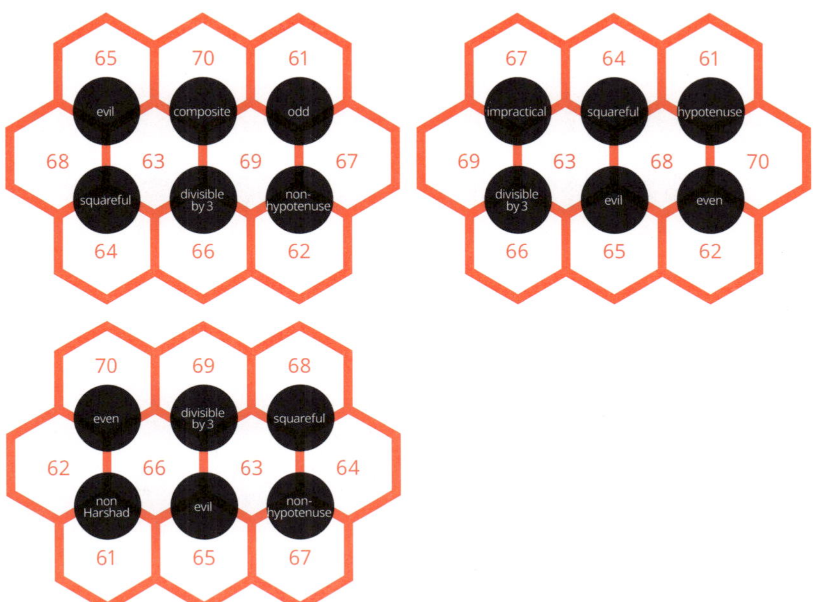

65

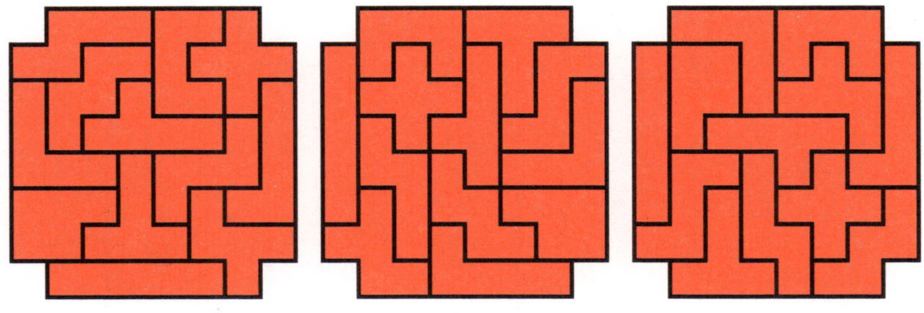

64

16	−	14	−	11	+	15
−		÷		×		÷
3	×	7	−	10	−	5
−		×		−		+
6	+	4	+	8	−	12
−		−		×		−
1	×	2	+	13	−	9

8	+	1	−	16	+	13
−		×		÷		−
4	÷	5	÷	2	×	15
×		+		−		−
7	−	10	+	12	−	3
÷		−		÷		+
14	+	9	−	6	−	11

15	÷	6	÷	5	×	12
−		×		−		+
7	−	2	×	4	÷	8
−		−		×		−
11	−	16	−	3	+	14
+		+		+		×
9	+	10	−	13	×	1

63

Going clockwise starting with the front position:
#63, #53, #33, #23, #73, #43, #93, #13, #3, #83

62

→	**1**	**+**	5	+	1
	−	**6**	−	7	−
	9	**+**	**1**	**×**	**7**
	−	**6**	**×**	6	**=**
	4	**+**	**7**	**=**	**62**

→	**1**	**+**	**5**	**+**	1
	−	**6**	**−**	**7**	−
	9	**+**	**1**	**×**	**7**
	−	**6**	**×**	6	**=**
	4	**+**	**7**	=	**62**

→	**1**	**+**	**5**	**+**	1
	−	**6**	**−**	**7**	−
	9	**+**	**1**	**×**	**7**
	−	**6**	**×**	6	**=**
	4	**+**	7	=	**62**

→	**7**	**+**	**9**	−	1
	×	**6**	**+**	4	×
	5	**×**	8	−	1
	+	**8**	−	8	=
	7	**−**	**9**	**=**	**62**

→	**7**	**+**	**9**	−	1
	×	6	**+**	4	×
	5	**×**	**8**	**−**	**1**
	+	**8**	**−**	**8**	**=**
	7	**−**	**9**	=	**62**

→	**7**	**+**	9	**−**	**1**
	×	**6**	**+**	**4**	**×**
	5	**×**	**8**	**−**	**1**
	+	**8**	**−**	**8**	**=**
	7	**−**	**9**	=	**62**

→	**7**	**+**	**8**	−	6
	−	**6**	**×**	8	×
	2	**+**	1	−	8
	+	**2**	−	5	=
	1	**+**	**3**	**=**	**62**

→	**7**	**+**	**8**	−	6
	−	**6**	**×**	**8**	×
	2	+	1	**−**	8
	+	**2**	**−**	**5**	=
	1	**+**	**3**	**=**	**62**

→	**7**	+	8	−	6
	−	**6**	**×**	**8**	**×**
	2	**+**	**1**	**−**	**8**
	+	**2**	**−**	**5**	=
	1	**+**	**3**	**=**	**62**

61

I ♥ cuban primes

For 61, $(5^3 - 4^3)/(5 - 4) = (125 - 64)/1 = 61$. Writing this as (61, 5, 4) to save time our other solutions are (7, 2, 1), (19, 3, 2), (37, 4, 3) all generated by the first equation. The case of 13 involves the second equation where $x = 3$ and $y = 1$.

This is our toughest alphametic yet, even with the hint. You will have various possibilities at certain stages and will have to keep those options open in your mind as you explore further to see which hold up.

The values of 2, 6 and 9 are already taken, leaving possible values of 0, 1, 3, 4, 5, 7 and 8. Looking at the units column 6 + 3 × N = C0 for some C. Only N = 8, C = 3 is possible. Carrying the 3 into the tens column we get 13 + 3 × E = C8 so E = 5 and we carry C = 2. In the hundreds column we now get 23 + V = CO which has 4 possible solutions (V, O, C) = (0, 3, 2), (1, 4, 2), (4, 7, 2) or (7, 0, 3).

Checking what happens if we carry a 3, the thousands column gives us 16 + 2 × T = CY where T and Y could take values 0, 1, 3, 4 or 7. Y would have to be even here and neither Y = 0 or Y = 4 work out, so we can exclude V = 7, O = 0 from the previous choices.

We still have 3 options here, so let's park that bit there for a moment and see what else we can find.

We know we carry a 2 into the thousands column we now get 15 + 2 × T = CY from possible values 0, 1, 3, 4 and 7. Possibilities are T = 1, Y = 7, C = 1; and T = 3, Y = 1 or T = 4, Y = 3 both with C = 2. If we take the first option the tens of thousands column would then give us 1 + L + 6 + 5 = C1 but this gives L = 9 and we would carry a new C = 2. This does not work in the hundreds of thousands column (you can check this). So we can exclude that first option from the thousands column.

So at least we know we are carrying a 2 into the tens of thousands column. Here we find 13 + L = CT (again for new value of C which is the digit we carry) where T must equal 3 or 4. We know only 0, 1, 3, 4 and 7 are options and we get L = 0, T = 3 and L = 1, T = 4 as possibilities.

So combining the above 2 paragraphs we know (T, Y, L) must be (3, 1, 0) or (4, 3, 1).

But from earlier (V, O) can equal (0, 3), (1, 4) or (4, 7). But these 5 letters can't all be 0, 1, 3 or 4, there are too many of them. So V = 4 and O = 7 must be correct. So T = 3, Y = 1, L = 0 also must be correct.

We have reached our solution:

E = 5, I = 2, L = 0, N = 8, O = 7, S = 9, T = 3, V = 4, X = 6, Y = 1

```
          926
         8285
       505458
      9263558
  +  82853558
  -----------
     92631785
```

60

Why you good-fer-somethin', two-bit, 60-diagonalled Platonic solid ...

I hope you didn't get too much of a headache trying to picture these complicated Platonic solids nested inside a sphere. It turns out the dodecahedron fits better, because it takes up 66.5% of the sphere, whereas the icosahedron can only manage 60.5%.

You might have thought that the icosahedron, having 12 faces, would push out further into the sphere than the 12 faces of the dodecahedron. But it turns out that what really matters is the number of vertices for each shape. If you imagine each solid starting out as a point in the middle of a sphere and expanding as a series of points until each one hits the edge of the sphere, the parts of the shape that travel farthest are the vertices.

The 20 vertices of the dodecahedron basically 'stretch' into more of the sphere than the 12 that the icosahedron has to offer.

$(9+7-1)\times((8-4)\div(5-3)+2)=60$

$(9+3)\times((7-5)\times(8-4-2)+1)=60$

$(7\times((9+3)\div(8+4)+1)-2)\times 5=60$

$(((7-2)\times 5+8)\times 9+3)\div(4+1)=60$

$(9+1)\times(7-4)\times(8\div(5-3)-2)=60$

59

6		1		2
(14)		(6)		
8		5		7
(11)		(45)		(3)
3		9	(5)	4

6	(15)	9	(72)	8
7	(8)	1	(5)	5
				(20)
2		3	(12)	4

3		6		1
(2)				(7)
5		4		7
(7)		(5)		(56)
2		9	(17)	8

Fun with 59

101	5	71
29	59	89
47	113	17

Further fun with 59

13# + 1 is the first composite Euclid number and factorises like this:

13# + 1 = 13 × 11 × 7 × 5 × 3 × 2 + 1 = 59 × 509

58

3	0	0	3	3	1	2	1	1	3	3	3	2	2	2	2	2	1
1	0	0	1	1	2	3	1	1	3	3	0	1	2	2	0	0	2
1	0	0	1	1	2	3	1	1	3	3	0	1	2	2	0	0	2
3	3	3	2	2	2	3	3	3	0	0	0	3	3	3	1	1	2
3	3	3	2	2	2	3	3	3	0	0	0	3	3	3	1	1	2
3	2	2	0	0	1	1	0	0	1	1	2	3	1	1	1	1	0

57

9	×	3	×	2	+	4	−	1	= 57
9	×	5	+	6	+	3	×	2	= 57
9	×	5	÷	3	+	7	×	6	= 57
7	×	4	×	2	+	3	÷	3	= 57
9	×	7	−	9	÷	3	×	2	= 57

56

Latin squares

a	b	c	d
b	a	d	c
c	d	a	b
d	c	b	a

a	b	c	d
b	d	a	c
c	a	d	b
d	c	b	a

a	b	c	d
b	a	d	c
c	d	b	a
d	c	a	b

a	b	c	d
b	c	d	a
c	d	a	b
d	a	b	c

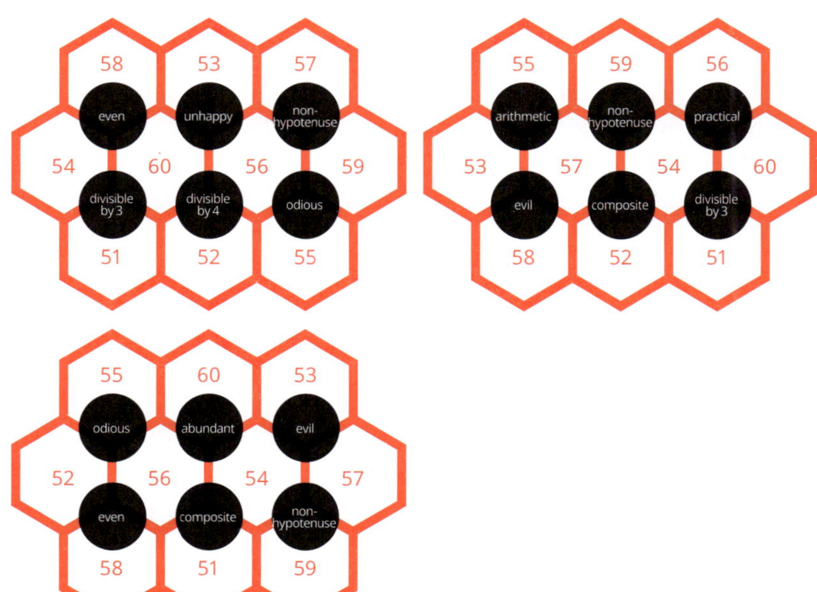

55

Graphic content

K_2 has one edge. K_3 has 3. K_4 has 6. K_5 has 10. The pattern would continue 1, 3, 6, 10, 15, 21, 28, 36, 45, 55; so K_{11} would have 55 edges.

This makes sense. You can look at it as linking up each of the 11 points with the other 10 points in the graph. This would give you 11×10 edges. But doing this would give you every edge twice. So we halve this number to get $11 \times 5 = 55$ edges for K_{11}. Congratulations, you've learnt some elementary graph theory!

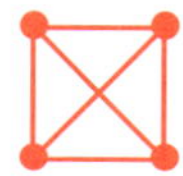

54

Tasty ...

As mentioned earlier, there are many possible solutions here. I've included some from the excellent *Geometry Snacks* but if you would like to see others, buy the book or search Alex Bellos's wonderful puzzle pages for *The Guardian*.

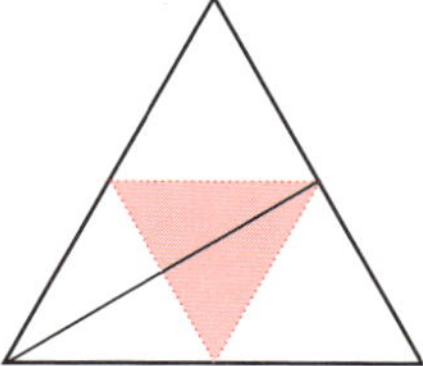

For the first problem, the diagram above clearly shows you that the shaded area is $1/4$ of the original figure. Hence the area is $54/4 = 13.5\text{ cm}^2$. Easy!

In order to solve this one you need to be familiar with Pythagoras's Theorem, which states that for right-angled triangles the square of the hypotenuse is equal to the sum of the squares on the other two sides.

Fifty-four is a big ugly number to use here so let's pretend the radius of the circles was 1 unit. Look at the triangle in the left image above. It has hypotenuse 2 and one of the other sides is 1. If the third side has length x, Pythagoras's Theorem tells us that the big square thus has side-length $1 + \sqrt{3}$, and area $(1 + \sqrt{3})^2$. But our radius was 54 cm not 1, so the area is $54^2 \times (1 + \sqrt{3})^2 = 11{,}664 + 5832\sqrt{3}$ cm^2.

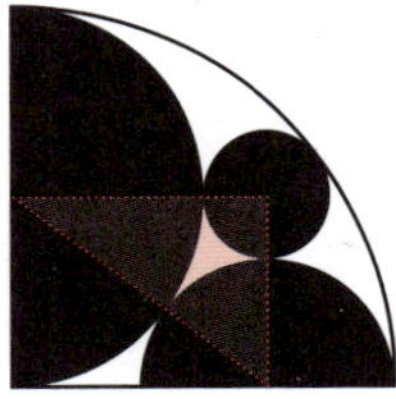

The smallest circle has radius 9, and the smaller semicircle has radius 18.

Firstly, some notation and basic deductions. Let the smallest circle have radius r, and the smaller semicircle radius R. Since the quarter-circle has radius 54, the radius of the larger semicircle must be 27.

This is just one way to do this. But to be honest, if you come close to getting it, well done you. Consider the right-angled triangle that we have drawn joining the centres of the relevant circles and semicircles. The hypotenuse is $27 + R$, and the sides have length 27, and $r + 27$. So, from Pythagoras:

$$27^2 + (r + 27)^2 = (27 + R)^2$$

We also see that the shorter side of the triangle is the same as the radius of the larger semicircle, or that $r + R = 27$, or $r = 27 - R$. So

$$27^2 + (54 - R)^2 = (27 + R)^2.$$

You don't need me to tell you this looks a bit rough. But it's really just expanding some brackets and if you've followed us this far I'm assuming you can do this. Upon expansion you get $54^2 = 162R$ or $R = 18$. And then $r = 27 - R = 9$.

Cubetastic

$7^3 = 343$, $54^3 = 157{,}464$ and $57^3 = 185{,}193$. Well done if you got all three of these correct. If you didn't, you'll struggle to get our final answer. But if you got these three tricky cubes correct you've done most of the heavy lifting. Adding them together we get 343,000. Noting what you'd already calculate that $7^3 = 343$, it's obvious that the sum of these three cubes is 70^3.

9	÷	3	−	13	+	15
+		−		−		−
14	−	2	+	5	−	12
−		+		×		−
11	+	4	−	16	+	6
−		×		÷		+
7	÷	1	−	10	+	8

15	−	13	÷	1	+	3
−		×		×		−
12	−	5	−	16	+	14
−		−		−		+
9	÷	6	+	7	÷	2
+		×		−		×
11	−	10	−	4	+	8

16	−	14	−	7	+	10
−		÷		+		−
3	+	1	×	11	−	9
+		+		−		×
4	×	6	÷	8	+	2
−		−		−		+
12	−	15	−	5	+	13

53

Going clockwise starting with the front position:
#53, #43, #73, #33, #3, #13, #23, #93, #63, #83

52

6	×	9	×	9
+	5	−	6	−
2	+	4	×	2
−	2	×	8	=
1	−	2	=	52

6	×	9	×	9
+	5	−	6	−
2	+	4	×	2
−	2	×	8	=
1	−	2	=	52

6	×	9	×	9
+	5	−	6	−
2	+	4	×	2
−	2	×	8	=
1	−	2	=	52

9	+	9	−	2
+	2	×	6	+
8	×	8	×	9
−	3	−	9	=
6	×	8	=	52

9	+	9	−	2
+	2	×	6	+
8	×	8	×	9
−	3	−	9	=
6	×	8	=	52

9	+	9	−	2
+	2	×	6	+
8	×	8	×	9
−	3	−	9	=
6	×	8	=	52

→

6	×	6	−	9
+	7	×	9	+
8	×	8	−	9
−	5	−	8	=
1	−	6	=	52

→

6	×	6	−	9
+	7	×	9	+
8	×	8	−	9
−	5	−	8	=
1	−	6	=	52

→

6	×	6	−	9
+	7	×	9	+
8	×	8	−	9
−	5	−	8	=
1	−	6	=	52

51

We have values T = 7, E = 2 and N = 6, so we only have 0, 1, 3, 4, 5, 8 and 9 available to use. We have our entire units and tens columns already so we carry a two into the hundreds column where we get 1 + R + I + 7 + 2 + 2 = CO. This gives 12 + R + I = CO. So C = 2 is the only option, so R and I must be 3 and 5 in some order. Carrying the 2 to the thousands column we get 2 + H + 6 + 7 + 7 = CY or 22 + H = CY. Only 0, 1, 4, 8 and 9 are left as options so H = 9, Y = 1 and we carry C = 3. Now the tens of thousands column gives us 3 + 7 + R + F = C7 where R = 3 or 5 and F can only be 0, 1, 4 or 8. This gives R = 3, F = 4 as the only possibility. So I = 5. The remaining values soon follow (I trust you by now) and you get the solution:

E = 2, F = 4, H = 9, I = 5, N = 6, O = 0, R = 3, T = 7, U = 8, Y = 1

```
      79322
       6562
        726
   40837226
+   4547226
-----------
   45471062
```

50

(9 − 4)×(8 − 3)×(6 − (7 + 1) ÷ 2) = 50

(9 − 4)×(7 − (3 + 1)÷(8 − 6))× 2 = 50

((9 + 7) ÷ 2 − 3)×(4 + 1)×(8 − 6) = 50

(8 + 6 − 4)×(7 − (9 − 3)÷(2 + 1)) = 50

((8 + 3)×(6 + 9 ÷ (7 − 4))+ 1) ÷ 2 = 50

Getting eban

The list of numbers is made up of 'ebans' — numbers in which the letter 'e' is 'banned'. So the next number which does not contain the letter 'e' is ... fifty.

49

6	(48)	8	(4)	2
		(1)		
5		9	(3)	3
7	(28)	4	(4)	1

8		4	(13)	9
(2)		(20)		(3)
6		5		3
(8)		(4)		
2		1		7

3		4		2
(18)		(1)		(4)
6	(11)	5		8
(3)		(12)		(8)
9		7		1

48

Waaaay back in '48

The prime factors of 48 are 2, 2, 2, 2 and 3, so the term $+\frac{1}{48}$ occurs.
The prime factors of 25 are 5 and 5, so the term $+\frac{1}{25}$ occurs.
The prime factors of 70 are 2, 5 and 7, so the term $-\frac{1}{70}$ occurs.
The prime factors of 2379 are 3, 13 and 61, and 61 is of the form $4 \times 15 + 1$, so the term $-\frac{1}{61}$ occurs in the series as does the term $+\frac{1}{2379}$.

You down with CRT?

1) *Not* 10c. If that were the case the bat would cost $1.10 and the total would be $1.20. The correct answer is 5c (the bat costs $1.05).

2) *Not* 100 minutes. The 5 machines make 5 widgets between them in 5 minutes. So each individual machine makes one widget in 5 minutes. So 100 machines would make 100 widgets ... in 5 minutes.

3) *Not* 24 days. If the patch doubles in size every day, it was half as big one day before. So it covered half the lake after 47 days.

1	0	0	0	0	2
3	2	2	2	2	0
3	2	2	2	2	0
3	1	1	3	3	0
3	1	1	3	3	0
0	0	0	0	0	1

3	0	0	0	0	2
0	0	0	0	0	1
0	0	0	0	0	1
3	3	3	1	1	1
3	3	3	1	1	1
3	3	3	3	3	2

2	1	1	1	1	2
3	1	1	0	0	0
3	1	1	0	0	0
1	3	3	0	0	3
1	3	3	0	0	3
2	1	1	3	3	0

47

7	×	3	×	2	+	4	+	1	= 47
8	+	7	×	3	+	6	×	3	= 47
9	+	9	×	4	+	8	÷	4	= 47
8	×	6	−	8	÷	2	+	3	= 47
9	×	7	÷	3	×	2	+	5	= 47

A right-royal ripper of a riddle!

As explained by quizmaster Dr Tim Paulden:

The left ink blotch obscures the number 395, while the right ink blotch obscures the numbers 86637 and 15913.

Here is one approach to reach the answer. Consider firstly the top line, where all the numbers are visible. For each number N on the left-hand side, if we multiply together the 4 other numbers on the left-hand side, add one, and then divide by N, we obtain the corresponding number appearing on the right-hand side. To illustrate, for the first number (N = 2) we find that:

$$\frac{(3 \times 11 \times 23 \times 31 + 1)}{2} = 11{,}765.$$

Similarly, for the second number (N = 3), we find that:

$$\frac{(2 \times 11 \times 23 \times 31 + 1)}{3} = 5229 \text{ and so on.}$$

For this pattern to hold for the second line as well, the left blotch must obscure the number 395, since:

$$\frac{(3 \times 7 \times 47 \times 395 + 1)}{2} = 194{,}933.$$

The numbers obscured by the right blotch can then be calculated as:

$$\frac{(2 \times 7 \times 47 \times 395 + 1)}{2} = 86{,}637 \quad \text{and} \quad \frac{(2 \times 3 \times 47 \times 395 + 1)}{2} = 15{,}913.$$

The two sets of whole numbers given on the left – (2, 3, 11, 23, 31) and (2, 3, 7, 47, 395) – are very unusual, because the 5 right-hand-side numbers generated by the above process are also whole numbers. These are, in fact, the only two sets of 5 whole numbers with this property (unless some of the numbers on the left or right are allowed to be equal to 1). Larger sets that follow the same pattern are also possible – for instance, (2, 3, 7, 47, 583, 1223) is such a 6-number set.

As several entrants noted, the reference to 'Stefan' in the question text is a nod to Štefan Znám, who studied sets of numbers with this property. (For further information, search online for 'Znám's problem'.)

46

All power to you

Don't forget that $2^{10} = 1024$ which is almost exactly 1000. When you move up by 10 powers of two you're multiplying the previous number by 1024. So from 2^{46} to 2^{56} we go from 14 to 17 digits and the first two digits increase from 70 to 72. Look at the two lead digits as we move up the powers on the list 70 (14 digits), 72 (17), 73 (20), 75 (23), 77 (26), 79 (29) before finally we kick over into a 32-digit number that begins with an 8.

Find a table of powers of 2 online and check out:

$2^5 = 32$
$2^{15} = 32,768$
$2^{25} = 33,554,432$ (quite possibly my favourite power of 2)
...
$2^{95} = 39,614,081,257,132,168,796,771,975,168$

Only at 2^{105} do we tick over to a number that begins with 4.

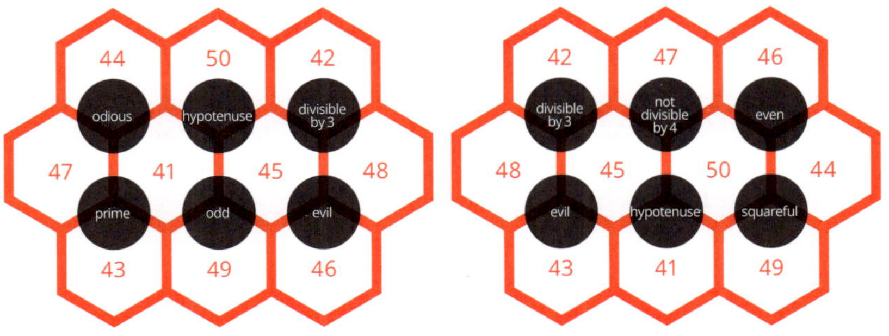

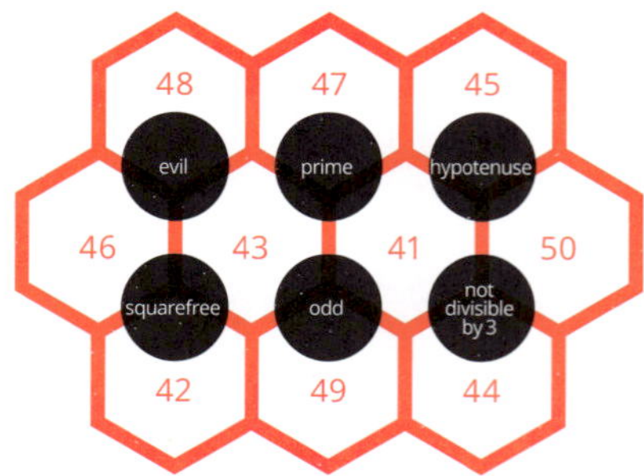

45

44

4	−	7	+	14	÷	2
+		+		+		×
9	+	1	−	11	+	5
−		+		−		÷
15	+	12	−	13	−	10
+		−		−		+
6	+	16	÷	8	÷	3

11	+	3	−	10	÷	1
−		×		−		×
5	×	2	+	8	−	14
+		−		÷		−
7	−	15	+	16	−	4
−		+		×		−
9	+	13	−	12	−	6

1	+	9	+	7	−	13
×		+		−		+
3	÷	15	×	2	×	10
÷		−		−		−
12	−	14	−	5	+	11
×		−		+		−
16	−	6	÷	4	×	8

43

Shine bright like Diophantine

$19^5 + 43^5 + 46^5 + 47^5 + 67^5 = 72^5$

$21^5 + 23^5 + 37^5 + 79^5 + 84^5 = 94^5$

$7^5 + 43^5 + 57^5 + 80^5 + 100^5 = 107^5$

$78^5 + 120^5 + 191^5 + 259^5 + 347^5 = 365^5$

$79^5 + 202^5 + 258^5 + 261^5 + 395^5 = 415^5$

$4^5 + 10^5 + 20^5 + 28^5 = 3^5 + 29^5$
$5^5 + 13^5 + 25^5 + 37^5 = 12^5 + 38^5$
$26^5 + 29^5 + 35^5 + 50^5 = 28^5 + 52^5$
$5^5 + 25^5 + 62^5 + 63^5 = 61^5 + 64^5$
$6^5 + 50^5 + 53^5 + 82^5 = 16^5 + 85^5$

Going clockwise starting with the front position:
#43, #83, #63, #33, #23, #93, #53, #13, #73, #3

42

→	8	×	9	+	5
	×	7	×	6	+
	8	×	6	−	3
	×	8	−	4	=
	2	×	2	=	42

→	8	×	9	+	5
	×	7	×	6	+
	8	×	6	−	3
	×	8	−	4	=
	2	×	2	=	42

→	8	×	9	+	5
	×	7	×	6	+
	8	×	6	−	3
	×	8	−	4	=
	2	×	2	=	42

→	7	×	8	+	2
	−	7	×	4	×
	4	−	8	+	2
	×	3	×	7	=
	5	+	4	=	42

→	7	×	8	+	2
	−	7	×	4	×
	4	−	8	+	2
	×	3	×	7	=
	5	+	4	=	42

→	7	×	8	+	2
	−	7	×	4	×
	4	−	8	+	2
	×	3	×	7	=
	5	+	4	=	42

→	9	+	5	+	5
	×	6	+	9	−
	1	+	2	−	8
	+	7	−	1	=
	7	−	3	=	42

→	9	+	5	+	5
	×	6	+	9	−
	1	+	2	−	8
	+	7	−	1	=
	7	−	3	=	42

→	9	+	5	+	5
	×	6	+	9	−
	1	+	2	−	8
	+	7	−	1	=
	7	−	3	=	42

Oh, brother!

This 66-word passage doesn't include a single repeated word.

Mathematician Mike Keith discovered it in 2002 searching for the longest such passage in the Project Gutenberg archives. You guessed it, this was his winner.

Interestingly, the same short story also includes the passage *'not the new boy. Why are you not the new boy? Why are you not.'* Which, somewhat unusually, comprises 15 3-letter words in a row.

41

You can be very proud of yourself if you deduced that:

E = 8, F = 1, N = 2, O = 9, R = 6, S = 7, T = 4, U = 5, V = 0, Y = 3

```
   19564882
        482
        482
+     78082
-----------
   19643928
```

If this proved too much for you, why not write out the alphametic again and give yourself the hint that TEN translates as 482. Can you deduce the rest?

PDF

$5 = 4 \times 1 + 1 = 2^2 + 1^2$. It's pretty easy to work out the n you require in the expression $4n + 1$, so for brevity let's write this as (5,2,1). Other such primes are (13,3,2), (17,4,1), (29,5,2), (37,6,1), (41,5,4), (53, 7,2), (61,6,5), (73, 8,3), (89,8,5), (97, 9,4).

40

$8 \times (3 + (6 - (5 - 1) \div 2) \div (9 - 7)) = 40$

$(7 + 1) \times ((9 - 5) \times (6 - 8 \div 2) - 3) = 40$

$(9 - 1) \times ((6 + (8 + 7) \div 5) \div 3 + 2) = 40$

$(9 - (8 - 5) \div (6 \div 3 + 1)) \times (7 - 2) = 40$

$(7 \times ((9 + 2) \div (8 + 3) + 1) - 6) \times 5 = 40$

39

2		8		6
(2)				(24)
1		3		4
(5)		(27)		(3)
5	(14)	9		7

2		6		3
(7)		(13)		
9		7		5
(9)		(28)		(3)
1		4	(32)	8

3	(12)	9		8
(3)				
1		4	(10)	6
(1)		(3)		
2		7	(35)	5

38

0	0	0	0	0	1
0	0	0	3	3	0
0	0	0	3	3	0
1	2	2	2	2	1
1	2	2	2	2	1
0	1	1	0	0	3

1	0	0	2	2	1
0	2	2	0	0	2
0	2	2	0	0	2
3	1	1	0	0	3
3	1	1	0	0	3
0	1	1	1	1	0

1	2	2	2	2	0
1	0	0	2	2	0
1	0	0	2	2	0
3	1	1	1	1	0
3	1	1	1	1	0
0	1	1	0	0	3

37

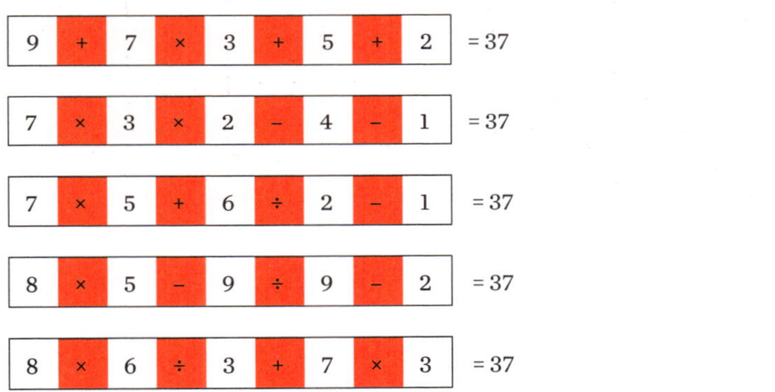

36

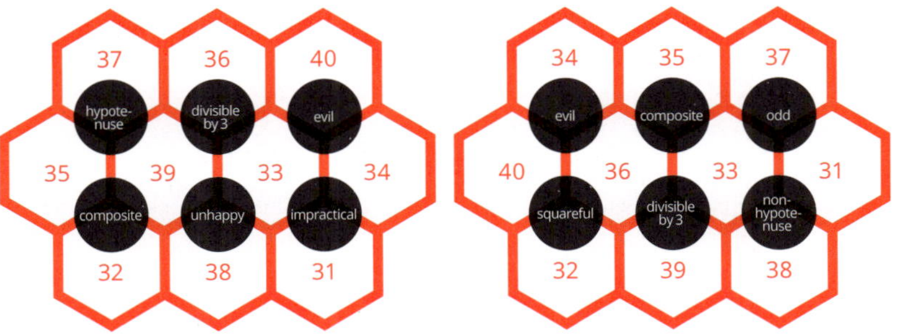

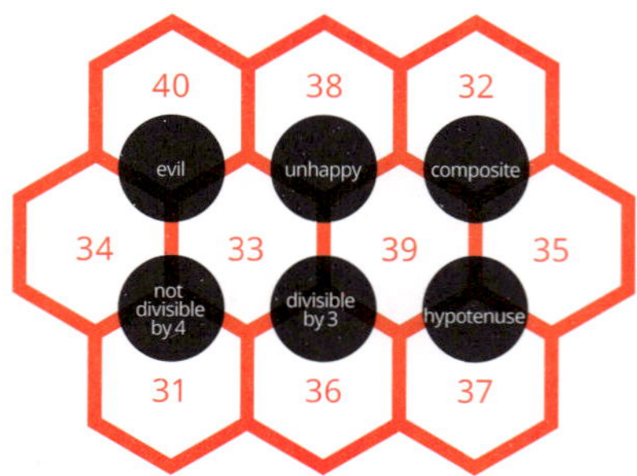

35

34

Fibbin' to 100

The Fibonacci numbers up to 100 are 1, 1, 2, 3, 5, 8, 13, 21, 34, 55 and 89. Starting with $F_1 = 1$ we get as far as $F_{11} = 89$. Now, 1 is relatively prime with every integer (technical but correct!) so F_1 is relatively prime with every F_n, and F_2 is relatively prime with every F_n when n is odd.

Next up is $F_3 = 2$ and 3 is relatively prime to 4, 5, 7, 8, 10 and 11. Well, $F_4 = 3$, $F_5 = 5$, $F_7 = 13$, $F_8 = 21$, $F_{10} = 55$ and $F_{11} = 89$ are all odd and therefore all relatively prime to $F_3 = 2$.

You would also need to compare F_4 to F_5, F_7, F_9 and F_{11} and so on. I leave that to you.

9	-	12	-	7	+	13
-		-		-		-
11	-	6	÷	2	-	5
-		÷		+		×
10	-	1	-	14	+	8
+		-		-		÷
15	-	3	×	16	÷	4

11	-	16	×	7	÷	14
-		÷		-		+
5	+	4	-	12	+	6
-		+		÷		-
13	+	8	-	3	-	15
+		-		÷		-
10	-	9	+	1	×	2

15	-	16	+	5	-	1
-		-		+		-
2	+	12	÷	3	÷	4
×		÷		-		+
10	-	6	-	14	+	13
+		-		+		-
8	+	11	-	9	-	7

33

Going clockwise starting with the front position:
#33, #3, #23, #43, #73, #83, #13, #63, #93, #53

32

→

5	–	5	×	7
×	3	+	6	×
6	+	1	+	3
×	8	×	8	=
1	×	9	=	32

→

5	–	5	×	7
×	3	+	6	×
6	+	1	+	3
×	8	×	8	=
1	×	9	=	32

→

5	–	5	×	7
×	3	+	6	×
6	+	1	+	3
×	8	×	8	=
1	×	9	=	32

→

9	×	9	–	6
–	9	–	5	+
7	–	1	×	2
–	3	×	3	=
1	×	2	=	32

→

9	×	9	–	6
–	9	–	5	+
7	–	1	×	2
–	3	×	3	=
1	×	2	=	32

→

9	×	9	–	6
–	9	–	5	+
7	–	1	×	2
–	3	×	3	=
1	×	2	=	32

→

5	–	8	–	2
×	3	×	6	–
6	×	7	×	1
×	2	+	6	=
7	×	6	=	32

→

5	–	8	–	2
×	3	×	6	–
6	×	7	×	1
×	2	+	6	=
7	×	6	=	32

→

5	–	8	–	2
×	3	×	6	–
6	×	7	×	1
×	2	+	6	=
7	×	6	=	32

31

E = 9, F = 1, I = 8, N = 4, O = 6, R = 7, T = 2, U = 0, W = 3, Y = 5

$$
\begin{array}{r}
236 \\
16072994 \\
1812994 \\
+\quad 239425 \\
\hline
18125649
\end{array}
$$

30

$6\times(4+(8-(9-5)\div2)\div(7-1))=30$

$(9+(8\div(7-5)+2)\div6)\times(4-1)=30$

$((6\div(8-5)+2)\times4-1)\times(9-7)=30$

$(8\div(9-7)+6\div(5-2))\times(4+1)=30$

$(8-1)\times(7-(4+2)\div(9-6))-5=30$

29

3		2		1
12		2		
9		4		7
15		20		15
6		5	13	8

5		3	21	7
14				15
9	15	6		8
10		2		
1		4		2

2		9	5	4
		27		
5	8	3		8
				2
1	6	7	42	6

28

0	0	0	2	2	2
1	1	1	0	0	3
1	1	1	0	0	3
2	0	0	0	0	2
2	0	0	0	0	2
1	0	0	0	0	1

0	0	0	0	0	2
1	0	0	0	0	2
1	0	0	0	0	2
2	1	1	1	1	0
2	1	1	1	1	0
1	3	3	0	0	1

1	3	3	2	2	0
2	0	0	0	0	0
2	0	0	0	0	0
0	1	1	0	0	0
0	1	1	0	0	0
2	1	1	1	1	3

27

6	×	3	+	4	×	2	+	1	= 37
9	×	6	+	8	–	7	×	5	= 37
7	×	6	÷	2	+	5	+	1	= 37
9	+	7	×	6	÷	3	+	4	= 37
7	×	4	–	8	÷	4	+	1	= 37

Oh snap!

Forfeit your shot — or even better, take a practice shot at the pole.

Huh? The important thing is that as long as both A and B are alive, neither of them has any incentive to shoot at you since you are the worst shot in the group (no offence). Your best bet is to wait until there is only one person left and then take your shot. Using your first shot for practice at the pole might even increase your chances despite your terrible aim (again, no offence).

Let's map this out. If you aim for A and kill him, B will aim at you with an 80% chance to kill you. If you aim for B and are successful, A will 100% certainly take you down.

If you miss, deliberately or otherwise, since B and A remain standing they will ignore you and target each other, because you're the less dangerous person each of them is up against (still no offence). Whether B kills A (80% chance) or B misses and A kills B (guaranteed), you will get your second shot at the sole survivor out of A and B. If you miss, you'll be in a fair bit of trouble, but missing the shot has only a 73% chance of happening.

The final odds are a bit more complicated to calculate because in theory you and B could continuously miss each other before one of you is finally successful, but this 73%-or-so chance of dying is better than the 100% and 80% on offer if you take your first shot and hit whomever you aimed at.

26

26 3 6 4 ... hut!

$26{,}364 = 26^3 \times 6/4$

In plain Randlish

You need to allow seven-letter words.

There are clearly 26 one-letter words. They are A B C ... X Y Z.

When we move to the two-letter words AA AB AC ... ZX ZY ZZ, there are 26 × 26 or 26^2 = 676 of them.

Similarly, there are 26^3 = 17,576 three-letter words (including CAT), 26^4 = 456,976 four-letter words, 26^5 = 11,881,376 five-letter words (including DDFFX) and 26^6 = 308,915,776 six-letter words.

But it's not until we add the 8,031,810,176 seven-letter words (for a total of 8,353,082,582 words) that we get more than the 7,632,819,325 people in the world of 2018.

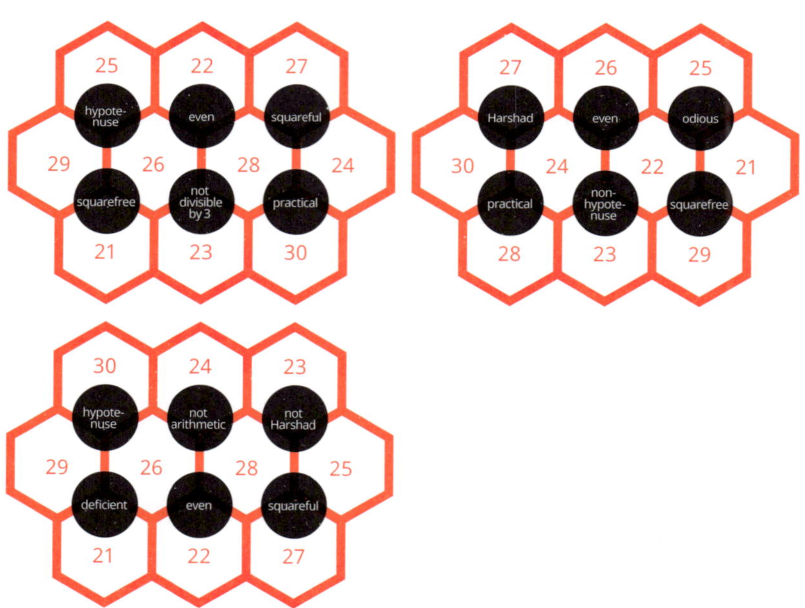

25

And in 25th place ...

The highly composite numbers up to 100 are 1, 2, 4, 6, 12, 24, 36, 48 and 60 which isn't superseded in number of factors until you hit 120.

MERRY XMAS TO ALL

There are two solutions, as demonstrated by L.A. Ringenberg of Eastern Illinois State College. A table of squares shows that ALL must be 100, 144, 400 or 900. If it's either of the first two, then XMAS has to be 2916 or 9216, and TO can't be a square. Ringenberg goes on to note that 'If ALL is 400, then XMAS has to be 1849, 3249, 6241 or 8649. Taking these four possibilities in turn:

- If ALL = 400 and XMAS = 1849, then M = 8, MERRY = 81225, and E and X now both represent 1, so that can't be right.
- If ALL = 400 and XMAS = 3249, then M = 2, MERRY = 27556, and TO = 81, so produces a valid solution: 27556 3249 81 400.
- If ALL = 400 and XMAS = 6241, then TO isn't a square.
- If ALL = 400 and XMAS = 8649, then M = 6 and L = 0 and there's no solution for MERRY.

Finally, going back to the possibility that ALL = 900, in this case XMAS = 1296 or 7396:

If ALL = 900 and XMAS = 1296, then TO isn't a square.
If ALL = 900 and XMAS = 7396, then M = 3, MERRY = 34225, and TO = 81, giving a second solution: 34225 7396 81 900'. Nice one.

24

Twenty-four ... more or less?

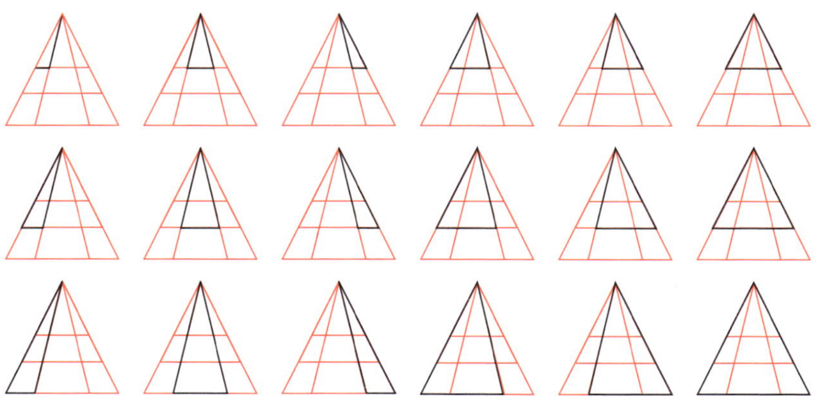

Observe that every triangle has to involve the top point of the big triangle. Let's think of all our triangles as starting from there and building out.

In addition, every triangle has to have a section of one of the 3 parallel lines as its base. So starting with the top parallel line there are 6 triangles we can form using various combinations of the downward lines.

But these 6 triangles just effectively 'repeat' on the other two parallel bases. This gives us 3 × 6 = 18 triangles as you can see here.

Twenty-four ... hours in a day

The trick to this problem was finding a way to make the most of the two 3s and two 8s that you could use. The answer is:

$$24 = \frac{8}{3 - (8/3)}$$

To see how this works, start by simplifying the bottom of the denominator. Here's what you get:

$$\frac{8}{3 - (8/3)} = \frac{8}{1/3} = 8 \times \frac{3}{1} = 24$$

Told you this was a toughie.

14	÷	7	×	3	−	4
−		+		+		−
13	−	9	+	10	−	12
−		−		−		+
15	÷	6	−	1	÷	2
+		−		×		×
16	−	8	−	11	+	5

6	×	3	−	9	−	7
×		−		−		+
2	−	4	+	12	−	8
+		+		+		+
5	+	13	−	16	÷	1
−		−		−		−
15	−	10	+	11	−	14

2	÷	12	×	6	+	1
+		−		−		÷
8	−	9	−	11	+	14
+		−		+		×
5	×	16	÷	10	÷	4
−		+		−		×
13	−	15	−	3	+	7

23

Going clockwise starting with the front position:
#23, #63, #33, #43, #83, #73, #3, #13, #93, #53

22

Regal rooms

No, there isn't. Try colouring the cells as shown below:

Any path that the king takes must alternate between dark and light rooms, so the king can succeed only if there are an equal number of each (or if their numbers differ by 1). In this diagram, there are 12 dark rooms and 10 light rooms, so there's no way to visit all of them on one connected tour, unless the path travels outside the palace.

A hex(agon) upon you

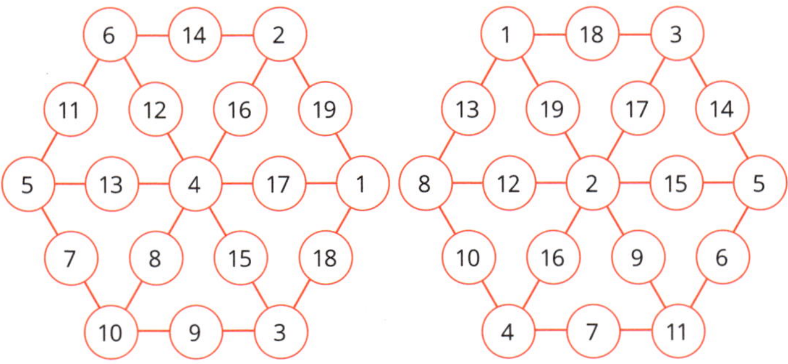

These two answers are provided by plus.maths.org. However, there are actually two other solutions to this puzzle. Did you find them?

If you're keen to look, read on for some fantastic insights from puzzle master Sean Gardiner to help narrow down your search because if you just go sticking numbers in at random, there are 19! = 121,645,100,408,832,000 possible combinations!

Firstly, note that the centre number can't be 10 or larger because there aren't 6 ways for two different available numbers to add up to 12 or less.

The centre number also has to be even because adding each line gives

$$22 \times 12 = 264 = (\text{sum of all the dots} = 190) + 2 \times (\text{all the corners}) + 5 \times (\text{centre})$$

since each corner gets counted a total of 3 times and the centre gets counted a total of 6 times. We can use the formula above to gain further insights. For example, if the centre is equal to 8, we get:

264 = 190 + 2 × (all the corners) + 5 × 8
264 = 230 + 2 × corners

So the corners sum to 17, which is impossible since the smallest possible corner sum is 1 + 2 + 3 + 4 + 5 + 6 = 21. This tells us we can't have an 8 in the centre.

We can try this logic again to see if we can place 6 in the centre. This time the corners would have to sum to exactly 22.

The only way this could happen would be if the corners were 1, 2, 3, 4, 5 and 7. But this would force 8 into one of the inside lines, since if it were on an outside line, the maximum value that line could take would be 5 + 7 + 8 = 20 < 22. If 8 is on an inside line, that line must also include the centre, 6. To sum to 22, it would need another 8 on the outside and this can't happen.

So we can't have a 6 in the centre either. Thus the only possible centres are 4, with the corners summing to 27; or 2, with the corners summing to 32.

In fact, I'll let you in on a secret here: the other two answers both have 2 at the centre.

Go on — try to find them now ...

How did you go?

I won't spoil it for you by putting the answers here, but they are easy to find online in the comments section of the original plus.maths.org article.

→	9	–	6	–	9
	+	5	+	3	–
	5	×	3	×	3
	×	6	×	8	=
	5	×	6	=	22

→	9	–	6	–	9
	+	5	+	3	–
	5	×	3	×	3
	×	6	×	8	=
	5	×	6	=	22

→	9	–	6	–	9
	+	5	+	3	–
	5	×	3	×	3
	×	6	×	8	=
	5	×	6	=	22

→	3	–	6	×	6
	–	5	×	9	+
	6	+	6	–	7
	–	4	–	6	=
	7	×	1	=	22

→	3	–	6	×	6
	–	5	×	9	+
	6	+	6	–	7
	–	4	–	6	=
	7	×	1	=	22

→	3	–	6	×	6
	–	5	×	9	+
	6	+	6	–	7
	–	4	–	6	=
	7	×	1	=	22

→	3	×	4	×	3
	–	7	×	5	×
	8	×	6	–	2
	–	3	–	3	=
	6	×	4	=	22

→	3	×	4	×	3
	–	7	×	5	×
	8	×	6	–	2
	–	3	–	3	=
	6	×	4	=	22

→	3	×	4	×	3
	–	7	×	5	×
	8	×	6	–	2
	–	3	–	3	=
	6	×	4	=	22

21

E = 5, G = 3, H = 7, I = 2, N = 8, O = 4, T = 1, W = 0, Y = 6

$$\begin{array}{r} 52371 \\ 52371 \\ 104 \\ 485 \\ +\quad 485 \\ \hline 105816 \end{array}$$

20

Here are some of my answers. There are others, of course.

$^{(10+5)}/_{5} + 1 = 4$
$^{10}/_{5} + 5 - 1 = 6$
$10 \times {}^{(5-1)}/_{5} = 8$
$10 \times {}^{5/5}/_{1} = 10$
$10 + 1 + {}^{5}/_{5} = 12$
$5 \times 5 - 10 - 1 = 14$
$5 \times 5 - 10 + 1 = 16$
$(10 + 5 + 5) \times 1 = 20$

And furthermore, what odd numbers can we get? (Thanks to my old mate Gareth White for 5 and 12.)

$5 + {}^{5}/_{10} \times 1 = 1$
$5 + {}^{1/10}/_{5} = 3$
$^{10}/_{(1+5/5)} = 5$
$5 + {}^{10}/_{5} \times 1 = 7$
$10 + 5 - 5 - 1 = 9$
$10 + 5 - 5 + 1 = 11$
$5 \times 5 - 10 \times 1 = 15$
$10 + 5 + 5 - 1 = 19$

According to my other old math-mate Sean Gardiner, 13, 17 and 18 are impossible. Allowing for square root signs you can get $10 + 5 + \sqrt{5}-1 = 17$ and $10 + 5 - \sqrt{5}-1 = 13$, but that's as close as you can get.

$((9+7)\times(6+5)\div(3-1)-8)\div 4 = 20$

$(9\times(8-1)-3)\div(5\times(6-4)-7) = 20$

$8\times(4+1)\div(6-(9-7)\times(5-3)) = 20$

$(7+3)\times(4-(9+5)\div(8+6)-1) = 20$

$(7\times((8+3)\div(6+5)+1)-9)\times 4 = 20$

Rock and (20c) roll!

The easier of the two answers is to the second question. The 20 on coin A will be written the same way whether you roll coin A over or under coin B. But the amazing bit is that the 20 will *be written upwards*.

As a fixed point on coin A is traced during its rotation around coin B, it traces out a figure called a 'cardioid'. You can see from the diagram below that once the coin has rolled to the top, which is halfway to the other side, the point on the right of the rolling coin is now on the left. Once it has rolled this far again, the dot will be back on the right-hand side where it started. The 20 on the coin is facing UP again! I know. The result is the same if the coin rotates over or under. The red dotted line that our starting point traces out is the cardioid.

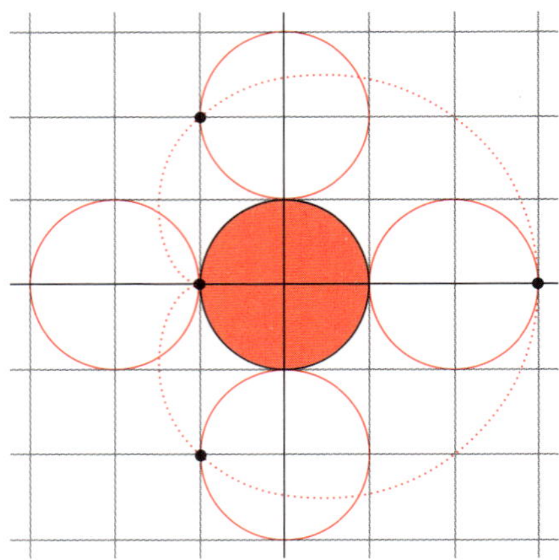

19

4		9	(1)	8
(20)		(9)		(6)
5		1		2
(12)				(6)
7		6	(2)	3

3	(18)	6	(48)	8
7	(2)	5		9
		(1)		
1	(3)	4	(8)	2

1		2		4
(5)				(11)
5		6	(1)	7
(4)		(14)		(4)
9		8	(5)	3

18

1	1	1	0	0	-1
-1	0	0	1	1	-1
-1	0	0	1	1	-1
2	1	1	2	2	1
2	1	1	2	2	1
0	-1	-1	-1	-1	2

0	-1	-1	-1	-1	0
-1	0	0	1	1	2
-1	0	0	1	1	2
0	2	2	2	2	1
0	2	2	2	2	1
0	-1	-1	-1	-1	2

1	-1	-1	0	0	1
0	0	0	-1	-1	2
0	0	0	-1	-1	2
2	2	2	0	0	1
2	2	2	0	0	1
1	1	1	1	1	-1

17

7	×	3	−	8	÷	6	×	3	= 17
9	+	8	÷	4	÷	2	+	7	= 17
7	−	9	+	5	×	3	+	4	= 17
9	×	7	−	8	×	5	−	6	= 17
9	−	7	×	4	+	6	×	6	= 17

16

17
15
18
unhappy
composite
Harshad
11
14
20
21
non-hypote-nuse
arithmetic
divisible by 4
19
13
16
16
20
19
not divisible by 3
hypote-nuse
prime
14
13
17
11
not divisible by 4
impractical
unhappy
18
15
12
19
18
15
non-hypote-nuse
not arithmetic
composite
11
12
16
14
unhappy
divisible by 4
odious
17
20
13

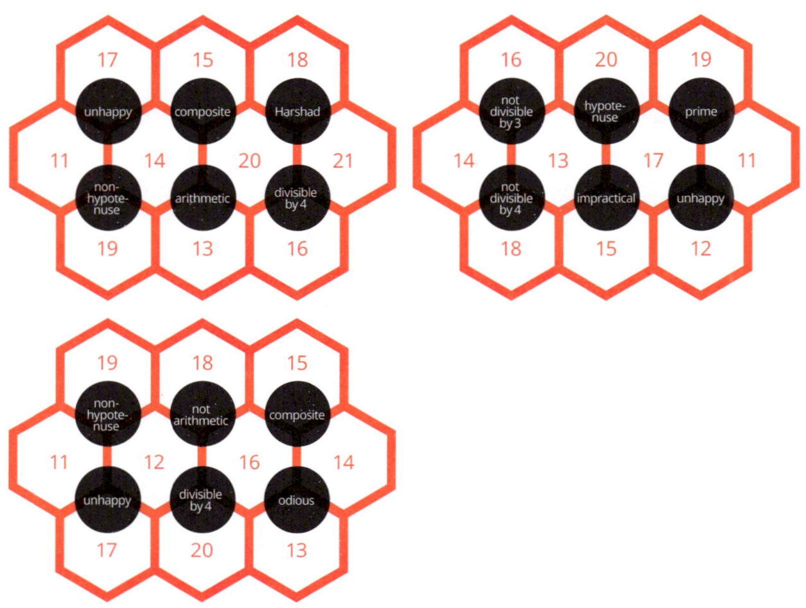

15

14

16	-	4	-	10	-	1
+		-		÷		-
12	÷	2	-	15	÷	3
-		×		÷		-
14	÷	7	-	6	+	5
-		+		×		+
13	-	11	-	9	+	8

15	-	7	÷	6	×	12
÷		-		+		+
10	÷	8	-	4	÷	16
-		-		-		-
1	-	11	+	14	-	3
÷		+		+		×
2	+	13	-	5	-	9

16	-	9	-	2	-	4
-		×		÷		-
13	-	5	+	1	-	8
-		÷		×		+
12	-	15	-	7	+	11
+		÷		÷		-
10	÷	3	-	14	÷	6

13

Going clockwise starting with the front position:
#13, #23, #73, #3, #93, #43, #63, #53, #33, #83

12

The hands of a clock ...

The hands are on each other at midnight when the minute hand and hour hand are both on the 12 position. They next coincide 'just after 1 am'; again after 2 am; after 3 am, 4 am, 5 am, 6 am, 7 am, 8 am, 9 am and 10 am. But the 'after 10 am' occurrence is very close to 11 am, so the 'after 11 am' event is actually precisely midday. So in the 12 hours from midnight until just before midday it has happened 11 times. It will happen another 11 times from midday until midnight. So from midnight until midnight, counting the starting midnight but not the finishing one (because they must be on separate days), we get 22 occurrences of the hands on top of each other.

When exactly does it first occur? It happens 11 times in 12 hours. Which is 11 times in 12 × 60 × 60 seconds (seeing I asked for seconds in the answer). This is 11 times every 43,200 seconds or once every 43,200 ÷ 11 = 3927.27 ... seconds. This is 3600 + (an hour) + 300 seconds (5 minutes) + 27.27 seconds. So the hands first lie on each other again at 1.05 am and 27.3 seconds.

→	**1**	**-**	**4**	-	7
	-	1	**+**	**5**	**×**
	2	+	8	-	**3**
	-	3	-	8	**=**
	1	-	1	=	**12**

→	**1**	-	4	-	7
	-	**1**	**+**	**5**	**×**
	2	**+**	8	-	**3**
	-	**3**	-	8	**=**
	1	-	1	=	**12**

→	**1**	**-**	4	-	7
	-	**1**	**+**	**5**	**×**
	2	**+**	**8**	**-**	**3**
	-	3	**-**	**8**	=
	1	-	**1**	**=**	**12**

→

3	–	9	+	2
–	2	×	5	–
2	–	7	–	3
×	4	–	1	=
3	+	7	=	12

→

3	–	9	+	2
–	2	×	5	–
2	–	7	–	3
×	4	–	1	=
3	+	7	=	12

→

3	–	9	+	2
–	2	×	5	–
2	–	7	–	3
×	4	–	1	=
3	+	7	=	12

→

6	+	2	×	9
×	7	×	3	–
9	+	9	+	3
×	4	–	1	=
4	+	2	=	12

→

6	+	2	×	9
×	7	×	3	–
9	+	9	+	3
×	4	–	1	=
4	+	2	=	12

→

6	+	2	×	9
×	7	×	3	–
9	+	9	+	3
×	4	–	1	=
4	+	2	=	12

11

And our solution is:

E = 1, H = 4, L = 7, N = 9, O = 3, R = 6, T = 8, V = 2, W = 0

$$
\begin{array}{r}
84611 \\
84611 \\
803 \\
803 \\
+ \quad 391 \\
\hline
171219
\end{array}
$$

10

Sweet geeky dreams ...

A calculator tells us the LHS ≈ 0.1303955989...

For the right-hand side:

$$\frac{1}{1^3 \times 2^3} = \frac{1}{8} = 0.125$$

$$\frac{1}{1^3 \times 2^3} + \frac{1}{2^3 \times 3^3} = \frac{1}{8} + \frac{1}{216} = 0.1296296296296...$$

$$\frac{1}{1^3 \times 2^3} + \frac{1}{2^3 \times 3^3} + \frac{1}{3^3 \times 4^3} = \frac{1}{8} + \frac{1}{216} + \frac{1}{1728} = 0.130208333333...$$

$$\frac{1}{1^3 \times 2^3} + \frac{1}{2^3 \times 3^3} + \frac{1}{3^3 \times 4^3} + \frac{1}{4^3 \times 5^3} = \frac{1}{8} + \frac{1}{216} + \frac{1}{1728} + \frac{1}{8000} = 0.1303333333...$$

So after 4 terms we agree to the first 4 decimal places.

$(9-4)\times(8-(7+2-6)\times(5-3)) = 10$

$((8\times(4+3)-6)\div(9-7)-5)\div 2 = 10$

$(9\times(4+3)-6\div(7-5))\div(8-2) = 10$

$(9-(7-5)\times(6-4))\times(8-3\times 2) = 10$

$(9\times 6-4)\div(3+(7+5)\div(8-2)) = 10$

The *Good Will Hunting* Problem

For trees of order 8 you're looking at:

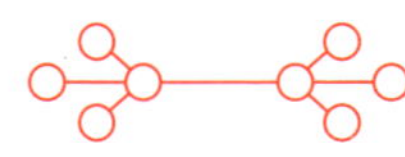

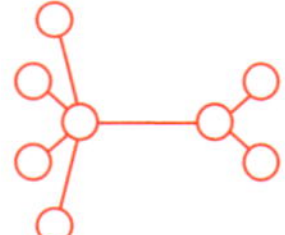

And for the *Good Will Hunting* Problem the 10 homeomorphically irreducible trees of order 10 are:

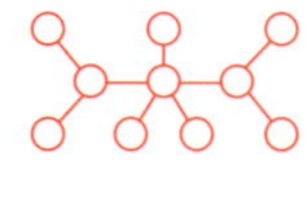
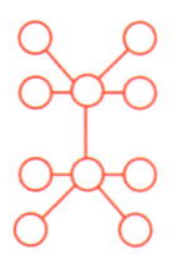
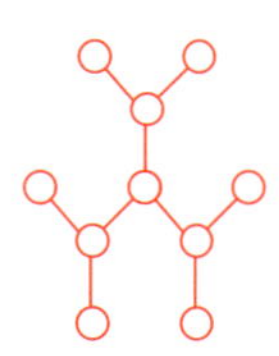

9

1	(9)	9	(1)	8
		(27)		
6	(2)	3		5
				(35)
4		2	(14)	7

1		8		9
(5)		(11)		(63)
5	(15)	3		7
				(9)
4		6	(12)	2

7	(15)	8	(12)	4
2		3		1
(8)				(9)
6	(1)	5	(14)	9

8

0	0	0	1	1	2	-1	-1	-1	2	2	1	1	1	1	2	2	-1
-1	-1	-1	0	0	2	0	1	1	1	1	2	-1	0	0	-1	-1	-1
-1	-1	-1	0	0	2	0	1	1	1	1	2	-1	0	0	-1	-1	-1
0	1	1	-1	-1	1	-1	-1	-1	-1	-1	2	-1	2	2	-1	-1	-1
0	1	1	-1	-1	1	-1	-1	-1	-1	-1	2	-1	2	2	-1	-1	-1
1	0	0	2	2	-1	-1	0	0	0	0	1	1	2	2	2	2	1

7

The (suit)case of the missing passports

Check suitcase 2, then 3, then 4. If they're not there, check suitcase 2, then 3, then 4 again. No matter what, you'll find the passports in time for your flight.

It should be obvious that randomly unpacking each suitcase won't guarantee you'll find the passports in time.

Let's start with an assumption, for argument's sake. Say the passports are in an even-numbered suitcase (so, suitcase 2 or 4). Let's check suitcase 2 first. If the passports are there, great. Have a nice flight! If not, we know they must be in suitcase 4, if we're assuming they were in an even-numbered suitcase.

Now, if the passports were in suitcase 4 on day one when you checked suitcase 2, then they must have been moved that evening to suitcase 3 or 5. So on day 2, check suitcase 3. Found the passports? Great! Bali here we come!

But what if they're not there? According to our assumption they must currently be in suitcase 5. Simply check suitcase 4 tomorrow (on day 3) and you're all done. Time to sit back and ruminate on the extraordinary achievement of having managed to pack 7 days early.

Of course there's one glaring oversight with this neat plan. What if the passports were in an *odd*-numbered suitcase, so 1, 3 or 5?

If this is the (suit)case (ha!) then on day 2, they must have moved to either suitcase 2 or 4.

Let's combine our two assumptions. In scenario one, checking suitcases 2, then 3, then 4 in the first 3 days, we know we'll find the passports if they started in an even-numbered suitcase. If you don't find them, then we know they *must* have started in an odd-numbered suitcase. So on day 4, if we still haven't found them, we can repeat the process since we now know that the passports must be in an even-numbered suitcase.

Bon voyage!

5	×	4	−	9	−	7	+	3	= 7
9	−	8	÷	4	×	3	+	4	= 7
9	×	7	÷	3	−	9	−	5	= 7
8	÷	7	×	6	+	1	÷	7	= 7
9	÷	6	+	8	−	5	÷	2	= 7

6

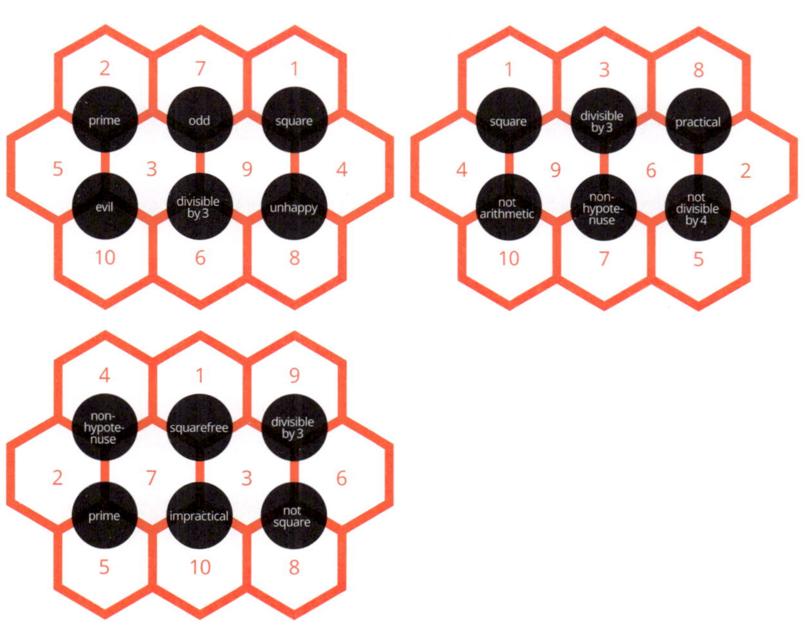

ANSWERS

5

4

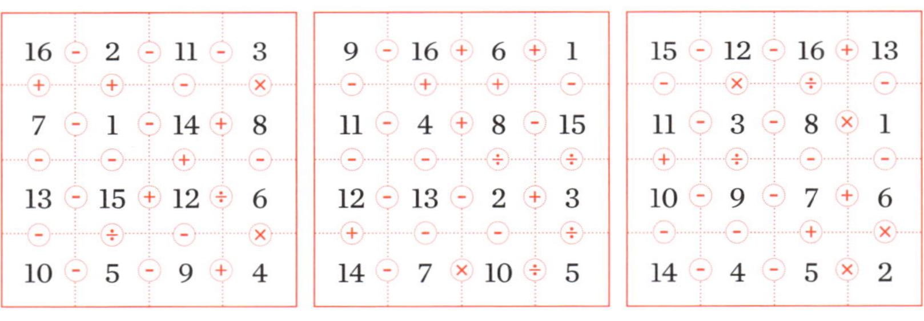

A maze

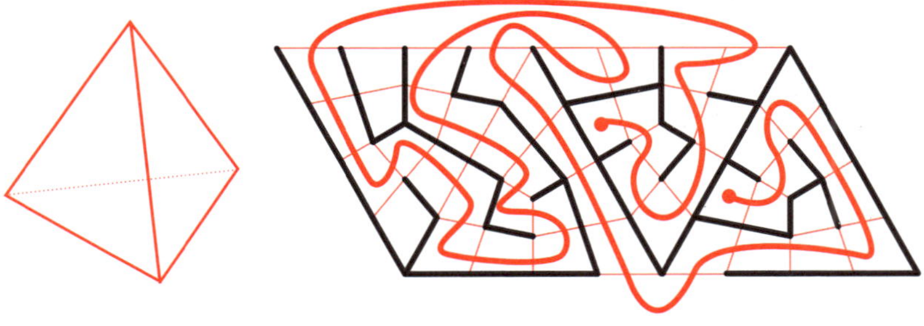

3

Going clockwise starting with the front position:
#3, #23, #73, #13, #83, #43, #53, #93, #33, #63

2

→	**7**	**+**	**5**	**-**	**2**
	+	3	-	8	**×**
	5	+	9	×	**5**
	-	9	×	6	**=**
	9	×	8	=	**2**

→	**7**	**+**	**5**	**-**	**2**
	+	**3**	**-**	**8**	**×**
	5	**+**	9	×	**5**
	-	9	×	6	**=**
	9	×	8	=	**2**

→	**7**	**+**	**5**	**-**	**2**
	+	**3**	**-**	**8**	**×**
	5	**+**	**9**	×	5
	-	**9**	**×**	6	=
	9	**×**	**8**	**=**	**2**

→	**6**	×	**3**	**×**	**5**
	-	4	**×**	8	**+**
	3	**×**	**1**	**+**	**6**
	+	**7**	**-**	**4**	**=**
	1	×	9	=	**2**

→	**6**	**×**	**3**	**×**	**5**
	-	**4**	**×**	**8**	**+**
	3	**×**	**1**	**+**	**6**
	+	**7**	**-**	4	**=**
	1	×	9	=	**2**

→	**6**	×	**3**	**×**	**5**
	-	**4**	**×**	**8**	**+**
	3	**×**	**1**	**+**	**6**
	+	7	**-**	4	**=**
	1	**×**	**9**	=	**2**

→	**8**	**×**	**3**	**-**	8
	×	6	**×**	**2**	+
	9	**×**	**1**	+	1
	×	**1**	**-**	3	=
	8	+	**4**	**=**	**2**

→	**8**	**×**	**3**	-	8
	×	**6**	**×**	**2**	**+**
	9	**×**	**1**	**+**	**1**
	×	**1**	**-**	**3**	=
	8	**+**	**4**	**=**	**2**

→	**8**	**×**	**3**	**-**	8
	×	**6**	**×**	**2**	+
	9	**×**	**1**	**+**	**1**
	×	1	**-**	**3**	**=**
	8	**+**	**4**	=	**2**

1

I can't stress how proud you should be of yourself if you got the following answer:

E = 1, F = 9, H = 6, I = 2, N = 4, O = 8, R = 7, T = 5, W = 3, Y = 0

```
  42415114
  56275114
     56711
       538
       538
       841
       841
+      841
----------
  98750538
```

Glossary of terms for our hex puzzles

It's a good exercise to try to find what numbers belong to each category yourself when doing the hexagon puzzles (and indeed if you're a return reader, you might have picked up a few of these terms already!).

That said, here's a useful list for checking.

A prime number is a whole number with exactly two factors: 1 and itself. ***A composite number*** is a whole number with more than two factors. The number 1 is neither prime nor composite.

The primes up to 100 are: 2, 3, 5, 7, 11, 13, 17, 19, 23, 29, 31, 37, 41, 43, 47, 53, 59, 61, 67, 71, 73, 79, 83, 89, 97

A squareful number is a number that is divisible by a perfect square greater than 1. For example, 18 is squareful because it is divisible by $9 = 3^2$.

A squarefree number is a number that is not divisible by any perfect square other than 1. For example, 30 is squarefree because none of its factors other than 1 is square: 1, 2, 3, 5, 6, 10, 15, 30.

The squareful numbers up to 100 are: 4, 8, 9, 12, 16, 18, 20, 24, 25, 27, 28, 32, 36, 40, 44, 45, 48, 49, 50, 52, 54, 56, 60, 63, 64, 68, 72, 75, 76, 80, 81, 84, 88, 90, 92, 96, 98, 99, 100

A perfect number equals the sum of its proper factors – that is, all its factors except itself. Both 6 and 28 are perfect, because $6 = 1 + 2 + 3$ and $28 = 1 + 2 + 4 + 7 + 14$.

A number is abundant if it is less than the sum of its proper factors. For example, 12 is abundant because $12 < 1 + 2 + 3 + 4 + 6$.

A number is deficient if it is more than the sum of its proper factors. For example, 8 is deficient because $8 > 1 + 2 + 4$.

The abundant numbers up to 100 are: 12, 18, 20, 24, 30, 36, 40, 42, 48, 54, 56, 60, 66, 70, 72, 78, 80, 84, 88, 90, 96, 100

A practical number has the interesting property that all positive numbers less than it can be expressed as the sum of some subset of its factors. For example, 8 is a practical number because using only the proper factors of 8 (1, 2 and 4), we can make all the numbers less than 8: 1, 2, 1 + 2, 4, 4 + 1, 4 + 2, 4 + 2 + 1.

If a number doesn't have this property, it's called an ***impractical number***. For example, 10 is an impractical number because it is not possible to express 4 or 9 as the sum of some subset of its factors (1, 2 and 5).

The practical numbers up to 100 are: 1, 2, 4, 6, 8, 12, 16, 18, 20, 24, 28, 30, 32, 36, 40, 42, 48, 54, 56, 60, 64, 66, 72, 78, 80, 84, 88, 90, 96, 100

An arithmetic number has the property that the average of all its factors (including itself) is a whole number. Recall that to get the average (arithmetic mean) of a set of numbers, you add them all up and divide the result by the number of things you just added.

For example, 15 is an arithmetic number because the average of its factors (1, 3, 5 and 15) is $(1 + 3 + 5 + 15)/4 = 24/4 = 6$, a whole number. It turns out that there are many more arithmetic numbers than non-arithmetic numbers in general.

The non-arithmetic numbers up to 100 are: 2, 4, 8, 9, 10, 12, 16, 18, 24, 25, 26, 28, 32, 34, 36, 40, 48, 50, 52, 58, 63, 64, 72, 74, 75, 76, 80, 81, 82, 84, 88, 90, 98, 100

A number is called ***a Harshad number*** if it is divisible by the sum of its digits. For example, 48 is a Harshad number because 4 + 8 = 12 is a factor of 48.

The Harshad numbers up to 100 are: 1, 2, 3, 4, 5, 6, 7, 8, 9, 10, 12, 18, 20, 21, 24, 27, 30, 36, 40, 42, 45, 48, 50, 54, 60, 63, 70, 72, 80, 81, 84, 90, 100

Back at number 89 we played a game of taking a number, adding the squares of its digits, and doing the same to this new number, repeating until we either reached 1 or a longer loop (which turned out to be 4, 16, 37, 58, 89, 145, 42, 20, 4 ...).

A happy number is one that eventually reaches 1 when you apply this process. For example, 82 is a happy number because 82 becomes $8^2 + 2^2 = 68$, which becomes $6^2 + 8^2 = 100$, which becomes $1^2 + 0^2 + 0^2 = 1$.

An unhappy number is one that never reaches 1 when you apply this process, which means it gets stuck in the loop. For example, 25 is an unhappy number because 25 becomes $2^2 + 5^2 = 29$, which becomes $2^2 + 9^2 = 85$, which will lead into the loop since its reverse, 58, is part of the loop.

The happy numbers up to 100 are: 1, 7, 10, 13, 19, 23, 28, 31, 32, 44, 49, 68, 70, 79, 82, 86, 91, 94, 97, 100

A hypotenuse number is one whose square can be written as the sum of two positive perfect squares. For example, 5 is a hypotenuse number because $5^2 = 25 = 3^2 + 4^2$.

A non-hypotenuse number is one whose square cannot be written as the sum of two positive perfect squares.

The hypotenuse numbers up to 100 are: 5, 10, 13, 15, 17, 20, 25, 26, 29, 30, 34, 35, 37, 39, 40, 41, 45, 50, 51, 52, 53, 55, 58, 60, 61, 65, 68, 70, 73, 74, 75, 78, 80, 82, 85, 87, 89, 90, 91, 95, 97, 100

At number 49 we looked at different number base systems, and at number 32 we looked at base 2, also known as binary.

An ***odious number*** is a number whose binary representation has an odd number of 1s. For example, 14 is odious because 14 in binary is 1110, which has three 1s.

And finally, an ***evil number*** is a number whose binary representation has an even number of 1s. For example, 9 is evil because 9 in binary is 1001, which has two 1s.

The evil numbers up to 100 are: 3, 5, 6, 9, 10, 12, 15, 17, 18, 20, 23, 24, 27, 29, 30, 33, 34, 36, 39, 40, 43, 45, 46, 48, 51, 53, 54, 57, 58, 60, 63, 65, 66, 68, 71, 72, 75, 77, 78, 80, 83, 85, 86, 89, 90, 92, 95, 96, 99

Also by Adam Spencer

Big Book of Numbers

World of Numbers

Time Machine

The Number Games

And for budding number nerds ...

Enormous Book of Numbers

Number Crunchers

12 Days of Christmas!

The Number Detective

Mini Book of Numbers